MICROBIAL METHODS IN MOLECULAR BIOLOGY AND BIOTECHNOLOGY

MICROBIAL METHODS IN MOLECULAR BIOLOGY AND BIOTECHNOLOGY

Dr. Manendra Kumar

RANDOM

RANDOM PUBLICATIONS

NEW DELHI - 110 002 (INDIA)

Microbial Methods in Molecular Biology and Biotechnology

ISBN 978-93-51116-94-3

Published in 2015 in India by

RANDOM PUBLICATIONS

4376-A/4B, Gali Murari Lal, Ansari Road
New Delhi-110 002
Phone: +9111-43580356, 23289044
E-mail: randomexports@gmail.com; sales@randompublications.com; info@randompublications.com

Reprinted 2026

Type Setting by: Friends Media, Delhi-110089

Preface

Microbes (or microorganisms) are organisms that are too small to be seen by the unaided eye. They include bacteria, fungi, protozoa, microalgae, and viruses. Microbes live in familiar settings such as soil, water, food, and animal intestines, as well as in more extreme settings such as rocks, glaciers, hot springs, and deep-sea vents. The wide variety of microbial habitats reflects an enormous diversity of biochemical and metabolic traits that have arisen by genetic variation and natural selection in microbial populations. Historically, humans have exploited some of this microbial diversity in the production of fermented foods such as bread, yogurt, and cheese. Some soil microbes release nitrogen that plants need for growth and emit gases that maintain the critical composition of the Earth's atmosphere. Other microbes challenge the food supply by causing yield-reducing diseases in food-producing plants and animals. In our bodies, different microbes help to digest food, ward off invasive organisms, and engage in skirmishes and pitched battles with the human immune system in the give-and-take of the natural disease process. A genome is the totality of genetic material in the DNA of a particular organism. Genomes differ greatly in size and sequence across different organisms. Obtaining the complete genome sequence of a microbe provides crucial information about its biology, but it is only the first step towards understanding a microbe's biological capabilities and modifying them, if needed, for agricultural purposes.

Microbial biotechnology, enabled by genome studies, will lead to breakthroughs such as improved vaccines and better disease-diagnostic tools, improved microbial agents for biological control of plant and animal pests, modifications of plant and animal pathogens for reduced virulence, development of new industrial catalysts and fermentation organisms, and development of new microbial agents for bioremediation of soil and water contaminated by agricultural runoff. Microbial genomics and microbial biotechnology research is critical for advances in food safety, food security, biotechnology, value-added products, human nutrition and functional foods, plant and animal protection, and furthering fundamental research in the agricultural sciences. Microbial biotechnology has a variety of useful applications in agriculture. Assessing and managing environmental risks from transgenic microorganisms is an important issue for which scientists have developed research needs and priorities. The mapping of microbial genomes is a key technology to understanding microorganisms and devising ways to improve their use in agricultural production, food safety, and bio-based chemicals. Micro-organisms are essential tools in biotechnology, biochemistry, genetics, and molecular biology. The yeasts and fission yeast are important model organisms in science, since they are simple eukaryotes that can be grown rapidly in large numbers and are easily manipulated.

This is an ideal textbook for courses in applied microbiology and biotechnology courses, this book will also serve as an invaluable overview of recent advances in this field for professional life scientists and for the diverse community of other professionals with interests in biotechnology.

I thank all members of my team who have helped in the preparation of the book. My special thanks go to "Random Publications" who have published the book.

— Dr. Manendra Kumar

Contents

Chapter 1

Introduction

Biomaterials can generally be produced either in nature or synthesized in the laboratory using a variety of chemical approaches utilizing metallic components or ceramics. They are often used and/or adapted for a medical application, and thus comprises whole or part of a living structure or biomedical device which performs, augments, or replaces a natural function. Such functions may be benign, like being used for a heart valve, or may be bioactive with a more interactive functionality such as hydroxy-apatite coated hip implants. Biomaterials are also used every day in dental applications, surgery, and drug delivery. E.G. A construct with impregnated pharmaceutical products can be placed into the body, which permits the prolonged release of a drug over an extended period of time. A biomaterial may also be an autograft, allograft or xenograft used as a transplant material.

Materials scientists are currently paying more and more attention to the process inorganic crystallization within a largely organic matrix of naturally occurring compounds. This process typically generally occurs at ambient temperature and pressure. Interestingly, the vital organisms through which these crystalline minerals form are capable of consistently producing intricately complex structures. Understanding the processes in which living organisms are capable of regulating the growth of crystalline minerals such as silica could lead to significant scientific advances and novel synthesis techniques for nanoscale composite materials — or nanocomposites.

Biomineralization

Biomineralization (e.g. silicification) is quite common in the biological world and occurs in bacteria, single-celled organisms, plants (e.g. petrified wood), and animals (invertebrates and vertebrates). Crystalline minerals formed in this type of environment often show exceptional mechanical properties (e.g. strength, hardness, fracture toughness) and tend to form hierarchical structures that exhibit microstructural order over a range of length or spatial scales. The minerals are typically crystallized from an environment that is undersaturated with respect to certain metallic elements such as silicon, calcium and [[phosphorus, which are readily oxidized under conditions of neutral pH and low temperature (0 - 40 degrees C). Formation of the mineral may occur either within or outside of the cell wall of an organism, and specific biochemical reactions for mineral deposition exist that include lipids, proteins and carbohydrates. The significance of the cellular machinery cannot be overemphasized, and it is with advances in experimental techniques in cellular biology and the capacity to mimic the biological environment that significant progress is currently being reported.

Examples include silicates in algae and diatoms, carbonates in invertebrates, and calcium phosphates and carbonates in vertebrates. These minerals often form structural features such as sea shells and the bone mineral in mammals and birds. Organisms have been producing mineralized skeletons for nearly 600 million years. The most common biominerals are the phosphate and carbonate salts of calcium that are used in conjunction with organic polymers such as collagen and chitin to give mechanical strength to bones and shells. Other examples include copper, iron and gold deposits involving bacteria.

Thus, most natural (or biological) materials are complex composites whose mechanical properties are often outstanding, considering the weak constituents from which they are assembled. These complex structures, which have risen from hundreds of million years of evolution, are inspiring materials scientists interested primarily in the design of novel materials with exceptional physical properties for high performance in adverse conditions. Their defining characteristics such as hierarchy, multifunctionality, and the capacity for self-healing, are currently being investigated. The basic building blocks begin with the 20 amino acids and proceed to polypeptides and polysaccharides.

These, in turn, compose the basic proteins, which are the primary constituents of the 'soft tissues' common to most biominerals. With well over 1000 proteins possible, current research emphasizes the use of collagen, chitin, keratin, and elastin. The 'hard' phases are often strengthened by crystalline minerals, which nucleate and grow in a biomediated environment that determines the size, shape and distribution of individual crystals. The most important silicate phases have been identified as hydroxyapatite, silica, and aragonite. Using the classification of Wegst and Ashby, the principal mechanical characteristics and structures of a number of biological ceramics, polymer composites, elastomers, and cellular materials have been recently characterized. Selected systems in each class are being investigated with emphasis on the relationship between their microstructure over a range of length scales and their mechanical response (esp. fracture toughness).

Thus, the crystallization of inorganic materials in nature generally occurs at ambient temperature and pressure. Yet the vital organisms through which these minerals form are capable of consistently producing extremely precise and complex structures. Understanding the processes in which living organisms control the growth of crystalline minerals such as silica could lead to significant advances in the field materials science, and open the door to novel synthesis techniques for nanoscale composite materials, or nanocomposites.

Abalone Shell

Recently, the nacre from the nautilus shell of the abalone has become one of the more intensively studied biological structures in materials science community. The highly ordered microscale aragonite tiles separated by thin nanoscale organic sheets along with a macrostructure of larger periodic growth bands form a hierarchical composite.

Early work showed that the overall composite consists of only 5 wt.% organic material—yet the work to fracture was increased by up to 3000 times over inorganic $CaCO_3$ crystals as a result of the intricate hierarchy of structural organization. High-resolution SEM observations have since been performed on the microstructure of nacre. Clearly visible in these images are the neatly stacked (or ordered) mineral tiles separated by thin organic sheets along with a macrostructure of larger periodic growth bands which collectively form a hierarchical composite structure.

The process of mineral formation following periods of growth interruption (growth bands) has been described by several authors. Flat pearl implantation as well as a new trepanning method have been used to observe the transitory phases of calcium carbonate which nucleate and grow during this process. An initial random nucleation of the aragonite polymorph is observed.

This nucleation and growth process is followed by a transition towards spherulitic growth. During this transition, the structural characteristics of the organism is determined by a combination of mechanical and chemical actions. About 6 weeks after implantation, a steady-state growth of aragonite tiles begins after shorter and more irregular tiles cover the outer surface of the spherulites. Finally, an organic scaffolding is observed during the steady-state growth of tiled aragonite.

In addition, observations of mineral growth following the deposition of these membranes confirm the presence of mineral bridges originating from subsurface tiles and extending through the organic matrix. Field emission scanning electron microscopy as well as TEM work on fractured deproteinated (deproteination) nacre have revealed the presence of mineral bridges existing between individual layers of tiles. These nanoscale bridges have been identified as the origins of the extreme fracture toughness of this biomaterial.

Recent joint collaboration at UC Santa Barbara and UC San Diego has produced striking results, including high resolution SEM images of the microstructure of the mother-of-pearl (or nacre) portion of the abalone shell, which exhibits the *highest mechanical strength and fracture toughness of any non-metallic substance known.*

Self-assembly

Self-assembly is the most common term in use in the modern scientific community to describe the spontaneous aggregation of particles (atoms, molecules, colloids, micelles, etc.) without the influence of any external forces. Large groups of such particles are known to assemble themselves into thermodynamically stable, structurally well-defined arrays, quite reminiscent of one of the 7 crystal systems found in metallurgy and mineralogy (e.g. face-centered cubic, body-centered cubic, etc.). The fundamental difference in equilibrium structure is in the spatial scale of the unit cell (or lattice parameter) in each particular case. Molecular self-assembly is found widely in biological systems and provides the basis of a wide variety of complex biological structures. This includes an emerging class of mechanically superior biomaterials based on microstructural features and designs found in nature. Thus, self-assembly is also emerging as a new strategy in chemical synthesis and nanotechnology. Molecular crystals, liquid crystals, colloids, micelles, emulsions, phase-separated polymers, thin films and self-assembled monolayers all represent examples of the types of highly ordered structures which are obtained using these techniques. The distinguishing feature of these methods is self-organization.

Structural Hierarchy

Nearly all materials could be seen as hierarchically structured, especially since the changes in spatial scale bring about different mechanisms of deformation and damage. However, in biological materials this hierarchical organization is inherent to the microstructure. One of the first examples of this, in the history of structural biology, is the early X-Ray scattering work on the hierarchical structure of hair and wool by Astbury and Woods. In bone, for example, collagen is the building block of the organic matrix—a triple helix with diameter of 1.5 nm. These tropocollagen molecules are intercalated with the mineral phase (hydroxyapatite, a calcium phosphate) forming fibrils that curl into helicoids of alternating directions. These "osteons" are the basic building blocks of bones, with the volume fraction distribution between organic and mineral phase being about 60/40. In another level of complexity, the hydroxyapatite crystals are platelets that have a diameter of approximately 70–100 nm and thickness of 1 nm. They originally nucleate at the gaps between collagen fibrils.

Similarly, the hierarchy of abalone shell begins at the nanolevel, with an organic layer having a thickness of 20–30 nm. This layer proceeds with single crystals of aragonite (a polymorph of CaCO3) consisting of "bricks" with dimensions of 0.5 and finishing with layers approximately 0.3 mm (mesostructure).

Crabs are arthropods whose carapace is made of a mineralized hard component (which exhibits brittle fracture) and a softer organic component composed primarily of chitin. The brittle component is arranged in a helical pattern. Each of these mineral 'rods' (1 μm diameter) contains chitin–protein fibrils with approximately 60 nm diameter. These fibrils are made of 3 nm diameter canals which link the interior and exterior of the shell.

Applications

Biomaterials are used in:

- Joint replacements
- Bone plates
- Bone cement
- Artificial ligaments and tendons
- Dental implants for tooth fixation
- Blood vessel prostheses
- Heart valves

- Skin repair devices (artificial tissue)
- Cochlear replacements
- Contact lenses
- Breast implants.

Biomaterials must be compatible with the body, and there are often issues of biocompatibility which must be resolved before a product can be placed on the market and used in a clinical setting. Because of this, biomaterials are usually subjected to the same requirements of those undergone by new drug therapies. All manufacturing companies are also required to ensure traceability of all of their products so that if a defective product is discovered, others in the same batch may be traced.

Heart Valves

In the United States, 45% of the 250,000 valve replacement procedures performed annually involve a mechanical valve implant. The most widely used valve is a bileaflet disc heart valve, or St. Jude valve. The mechanics involve two semicircular discs moving back and forth, with both allowing the flow of blood as well as the ability to form a seal against backflow. The valve is coated with pyrolytic carbon, and secured to the surrounding tissue with a mesh of woven fabric called Dacron™ (du Pont's trade name for polyethylene terephthalate). The mesh allows for the body's tissue to grow while incorporating the valve.

Skin Repair

Most of the time "artificial" tissue is grown from the patients own cells. However, when the damage is so extreme that it is impossible to use the patient's own cells, artificial tissue cells are grown. The difficulty is in finding a scaffold that the cells can grow and organize on. The characteristics of the scaffold must be that it is biocompatible, cells can adhere to the scaffold, mechanically strong and biodegradable. One successful scaffold is a copolymer of lactic acid and glycolic acid.

Compatibility

Biocompatibility is related to the behaviour of biomaterials in various environments under various chemical and physical conditions. The term may refer to specific properties of a material without specifying where or how the material is to be used. For example, a material may elicit little or no immune response in a given organism, and may or may not able to integrate with a particular cell type or tissue).

The ambiguity of the term reflects the ongoing development of insights into how biomaterials interact with the human body and eventually how those interactions determine the clinical success of a medical device (such as pacemaker or hip replacement. Modern medical devices and prostheses are often made of more than one material—so it might not always be sufficient to talk about the biocompatibility of a specific material.

Also, a material should not be toxic unless specifically engineered to be so—like "smart" drug delivery systems that target cancer cells and destroy them. Understanding of the anatomy and physiology of the action site is essential for a biomaterial to be effective. An additional factor is the dependence on specific anatomical sites of implantation. It is thus important, during design, to ensure that the implement will fit complementarily and have a beneficial effect with the specific anatomical area of action.

Biopolymers

Biopolymers are polymers produced by living organisms. Cellulose and starch, proteins and peptides, and DNA and RNA are all examples of biopolymers, in which the monomeric units, respectively, are sugars, amino acids, and nucleotides. Cellulose is both the most common biopolymer and the most common organic compound on Earth. About 33% of all plant matter is cellulose.

Some biopolymers are biodegradable. That is, they are broken down into CO_2 and water by microorganisms. In addition, some of these biodegradable biopolymers are compostable. That is, they can be put into an industrial composting process and will break down by 90% within 6 months. Biopolymers that do this can be marked with a 'compostable' symbol, under European Standard EN 13432 (2000). Packaging marked with this symbol can be put into industrial composting processes and will break down within 6 months (or less). An example of a compostable polymer is PLA film under 20 μm thick: films which are thicker than that do not qualify as compostable, even though they are biodegradable. A home composting logo may soon be established: this will enable consumers to dispose of packaging directly onto their own compost heap.

Metalloprotein

Metalloprotein is a generic term for a protein that contains a metal ion cofactor. Metalloproteins have many different functions in cells, such as enzymes, transport and storage proteins, and signal transduction proteins. Indeed, about one quarter to one third of all proteins require metals to carry out their functions. The metal ion is usually coordinated by nitrogen, oxygen or sulfur atoms belonging to amino acids in the polypeptide chain and/or a macrocyclic ligand incorporated into the protein. The presence of the metal ion allows metalloenzymes to perform functions such as redox reactions that cannot easily be performed by the limited set of functional groups found in amino acids.

Storage and Transport Metalloproteins

Oxygen Carriers

Hemoglobin, which is the principal oxygen carrier in humans has four sub-units in which the iron(II) ion is coordinated by the planar, macrocyclic ligand protoporphyrin IX (PIX) and the imidazole nitrogen atom of a histidine residue. The sixth coordination site contains a water molecule or a dioxygen molecule. myoglobin has only one such unit. The active site is located in an hydrophobic pocket. This is important as, without it, the iron(II) would be irreversibly oxidised to iron(III). The equilibrium constant for the formation of HbO_2 is such that oxygen is taken up or released depending on the partial pressure of oxygen in the lungs or in muscle. In hemoglobin the four sub-units show a cooperativity effect which allows for easy oxygen transfer from hemoglobin to myoglobin.

In both hemoglobin and myoglobin it is sometimes incorrectly stated that the oxygenated species contains iron(III). It is now known that the diamagnetic nature of these species is due to the fact that the iron(II) is in the low-spin state. In oxyhemoglobin the iron atom is located in the plane of the porphyrin ring, but in the paramagnetic deoxyhemoglobin the iron atom lies above the plane of the ring. The change in spin state is a cooperative effect of higher crystal field splitting and smaller ionic radius of Fe^{2+} in the oxy-moiety.

Hemerythrin is another iron-containing oxygen carrier. The oxygen binding site is a binuclear iron centre. The iron atoms are coordinated to the protein through the carboxylate side chains of a glutamate and aspartate and five histidine residues. The uptake of O_2 by hemerythrin is accompanied by two-electron oxidation of the reduced binuclear centre to produce bound peroxide (OOH-).

Hemocyanins carry oxygen in the blood of most molluscs, and some arthropods such as the horseshoe crab. They are second only to hemoglobin in biological popularity of use in oxygen transport. On oxygenation the two copper(I) atoms at the active site are oxidised to copper(II) and the dioxygen molecules is reduced to peroxide, O_2^{2-}.

Cytochromes

Oxidation and reduction reactions are not common in organic chemistry as few organic molecules can act as oxidizing or reducing agents. Iron(II), on the other hand, can easily be oxidized to iron(III). This functionality is used in cytochromes which function as electron-transfer vectors. The iron atom in most cytochromes is contained in a heme group.

The differences between those cytochromes lies in the different side-chains. For instance Cytochrome a has a heme a prosthetic group and cytochrome b has a heme b prosthetic group. These differences result in different Fe^{2+}/Fe^{3+} redox potentials such that various cytochromes are involved in the mitochondrial electron transport chain. Cytochrome P450 enzymes perform the function of inserting an oxygen atom into a C—H bond, an oxidation reaction.

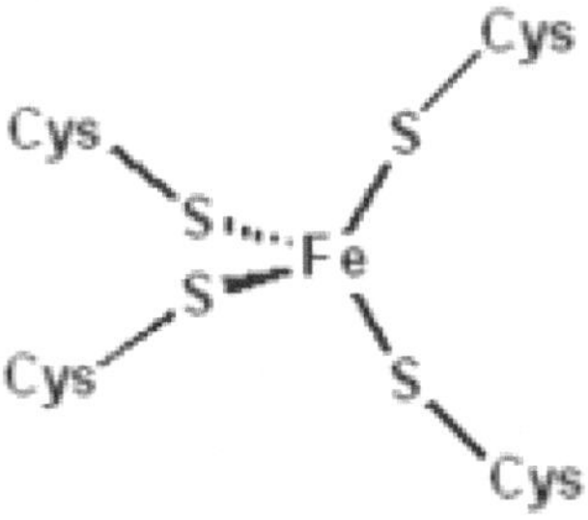

Rubredoxin

Rubredoxin is an electron-carrier found in sulfur-metabolizing bacteria and archaea. The active site contains an iron ion which is coordinated by the sulphur atoms of four cysteine residues forming an almost regular tetrahedron. Rubredoxins perform one-electron transfer processes. The oxidation state of the iron atom changes between the +2 and +3 states. In both oxidation states the metal is high spin, which helps to minimize structural changes.

Plastocyanin

Plastocyan is one of the family of blue copper proteins which are involved in electron transfer reactions. The copper binding site is described as a 'distorted trigonal pyramidal'. The trigonal plane of the pyramidal base is composed of two nitrogen atoms (N_1 and N_2) from separate histidines and a sulfur (S_1) from a cysteine. Sulfur (S_2) from an axial methionine forms the apex. The 'distortion' occurs in the bond lengths between the copper and sulfur ligands. The Cu-S_1 contact is shorter (207 picometers) than Cu-S_2 (282 pm). The elongated Cu-S_2

bonding destabilises the Cu^{II} form and increases the redox potential of the protein. The blue colour (597 nm peak absorption) is due to the Cu-S_1 bond where $S_{p\eth}$ to $Cu_{dx2\text{-}y2}$ charge transfer occurs.

In the reduced form of plastocyanin, His-87 will become protonated with a pK_a of 4.4. Protonation prevents it acting as a ligand and the copper site geometry becomes trigonal planar.

Iron Storage and Transfer

Iron is stored as iron(III) in ferritin. The exact nature of the binding site has not yet been determined. The iron appears to be present as an hydrolysis product such as FeO(OH). Iron is transported by transferrin whose binding site consists of two tyrosines, one aspartic acid and one histidine.

The human body has no mechanism for iron excretion. This can lead to iron-overload problems in patients treated with blood transfusions, as, for instance, with β-thallasemia.

Ceruloplasmin

Ceruloplasmin is the major copper-carrying protein in the blood. Ceruloplasmin exhibits oxidase activity, which is associated with possible oxidation of Fe^{2+} (ferrous iron) into Fe^{3+} (ferric iron), therefore assisting in its transport in the plasma in association with transferrin, which can only carry iron in the ferric state.

Metalloenzymes

Metalloenzymes all have one feature in common, namely, that the metal ion is bound to the protein with one labile coordination site. As with all enzymes, the shape of the active site is crucial. The metal ion is usually located in a pocket whose shape fits the substrate. The metal ion catalyzes reactions which are difficult to achieve in organic chemistry.

Carbonic Anhydrase

Active site of carbonic anhydrase. The three coordinating histidine residues are shown in pink, hydroxide in red and white, and the zinc in purple.

$$CO_2 + H_2O \rightleftharpoons H_2CO_3$$

This reaction is very slow in the absence of a catalyst, but quite fast in the presence of the hydroxide ion

$$CO_2 + OH^- \ HCO_3^-$$

A reaction similar to this is almost instantaneaous with carbonic anhydrase. The structure of the active site in carbonic anhydrases is well-known from a number of crystal structures. It consists of a zinc ion coordinated by three imidazole nitrogen atoms from three histidine units.

The fourth coordination site is occupied by a water molecule. The coordination sphere of the zinc ion is approximately tetrahedral. The positively charged zinc ion polarizes the coordinated water molecule and nucleophilic attack by the negatively charged hydoxide portion on carbon dioxide (carbonic anhydride) proceeds rapidly.

The catalytic cycle produces the bicarbonate ion and the hydrogen ion as the equilibrium"

$$H_2CO_3\ HCO_3^- + H^+$$

Favours dissociation of carbonic acid at biological pH values.

Vitamin B_{12}-dependent Enzymes

Vitamin B_{12} catalyzes the transfer of methyl ($-CH_3$) groups between two molecules, which involves the breaking of C-C bonds, a process that is energetically expensive in organic reactions.

The metal ion lowers the activation energy for the process by forming a transient Co-CH_3 bond. The structure of the coenzyme was famously determined by Dorothy Hodgkin and co-workers, for which she received a Nobel prize. It constists of a cobalt(II) ion coordinated by four nitrogen atoms of a corrin rings and a fifth Nitrogen atom from an imidazole group. In the resting state there is a Co—C σ bond with the 5' carbon atom of adenosine. This is a naturally occurring organometallic compound, which explains its function in trans-methylation reactions, such as the reaction carried out by methionine synthase.

Nitrogenase (Nitrogen Fixation)

The fixation of atmospheric nitrogen is a very energy-intensive process, as it involves breaking the very stable triple bond between the nitrogen atoms.

The enzyme nitrogenase is one of the few enzymes that can catalyse the process.

The enzyme occurs in certain bacteria. There are three components to its action: a molybdenum atom at the active site, Iron-sulfur clusters which are involved in transporting the electrons needed to reduce the nitrogen and an abundant energy source.

The energy is provided by a symbiotic relationship between the bacteria and a host plant, often a legume. The relationship is symbiotic because the plant supplies the energy by photosynthesis and benefits by obtaining the fixed nitrogen. The reaction may be written symbolically as

$$N_2 + 16\ MgATP + 8e^- \rightarrow 2NH_3 + 16\ MgADP + 16\ P_i + H_2$$

Where P_i stands for inorganic phosphate. The precise structure of the active site has been difficult to determine. It appears to contain a $MoFe_7S_8$ cluster which is able to bind the dinitrogen molecule and, presumably, enable the reduction process to begin.

The electrons are transported by the associated "P" cluster, which contains two cubical Fe_4S_4 clusters joined by sulphur bridges.

Superoxide Dismutase

The superoxide ion, O_2^- is generated in biological systems by reduction of molecular oxygen. It has an unpaired electron, so it behaves as a free radical. It is a powerful oxidising agent. These properties render the superoxide ion very toxic and are deployed to advantage by phagocytes to kill invading micro organisms.

Otherwise, the superoxide ion must be destroyed before it does unwanted damage in a cell. The superoxide dismutase enzymes perform this function very efficiently.

The formal oxidation state of the oxygen atoms is ½. In solutions at neutral pH, the superoxide ion disproportionates to molecular oxygen and hydrogen peroxide.

$$2\ O_2^- + 2\ H^+ \rightarrow O_2 + H_2O_2$$

In biology this type of reaction is call a dismutation reaction. It involves both oxidation and reduction of superoxide ions. The superoxide dismutase group of enzymes, abbreviated as SOD, increase the rate of reaction to near the diffusion limited rate. The key to the action of these enzymes is a metal ion with variable oxidation state which can act as either an oxidizing agent or as a reducing agent.

$$\text{Oxidation: } M^{(n+1)+} + O_2^- \rightarrow M^{n+} + O_2$$

$$\text{Reduction: } M^{n+} + O_2^- + 2H^+ \rightarrow M^{(n+1)+} + H_2O_2.$$

In human SOD the active metal is copper, as Cu^{2+} or Cu^+, coordinated tetrahedrally by four histidine residues. This enzyme also contains zinc ions. Other isozymes may contain iron, manganese or nickel. Ni-SOD is particularly interesting as it involves nickel(III), an unusual oxidation state for this element.

The active site Ni geometry cycles from square planar Ni(II), with thiolate (Cys2 and Cys6) and backbone nitrogen (His1 and Cys2) ligands, to square pyramidal Ni(III) with an added axial His1 side chain ligand.

Chlorophyll-containing Proteins

Chlorophyll plays a crucial role in photosynthesis. It contains a magnesium enclosed in a chlorin ring. However, the magnesium ion is not directly involved in the photosynthetic function and can be replaced by other divalent ions with little loss of activity. Rather, the photon is absorbed by the chlorin ring, whose electronic structure is well-adapted for its purpose.

Initially, the absorption of a photon causes an electron to be excited into a singlet state of the Q band. The excited state undergoes an intersystem crossing from the singlet state to a triplet state in which there are two electrons with parallel spin. This species is, effectively, a free radical, and is very reactive and allows an electron to be transferred to acceptors which are adjacent to the chlorophyll in the chloroplast. In the process chlorophyll is oxidized. Later in the phosynthetic cycle, chlorophyll is re-reduced. This reduction ultimately draws electrons from water, yielding molecular oxygen as a final oxidation product.

Signal-transduction Metalloproteins

Calmodulin is an example of a signal-transduction protein. It is a small protein which contains four EF-hand motifs, each of which can bind a Ca^{2+} ion. In an EF-hand loop the calcium ion is coordinated in a pentagonal bipyramidal configuration. Six Glutamic acid and Aspartic acid residues involved in the binding are in positions 1, 3, 5, 7, 9 of the polypeptide chain. At position 12 there is a glutamate or aspartate ligand which behaves as a (bidentate ligand), providing two oxygen atoms. The ninth residue in the loop is necessarily glycine due to the conformational requirements of the backbone. The coordination sphere of the calcium ion contains only carboxylate oxygen atoms and no nitrogen atoms. This is consistent with the hard nature of the calcium ion. The protein has two approximately symmetrical domains, separated by a flexible "hinge" region. Binding of calcium causes a conformational change to occur in the protein. Calmodulin participates in an intracellular signalling system by acting as a diffusible second messenger to the initial stimuli.

Transcription

Many transcription factors contain a structure known as a zinc finger, this is a structural module where a region of protein folds around a zinc ion. The zinc does not directly contact the DNA that these proteins bind to, instead the cofactor is essential for the stability of the tightly-folded protein chain. In these proteins the zinc ion is usually coordinated by pairs of cysteine and histidine side chains.

Other Metalloenzymes

Ion	Examples of enzymes containing this ion
Magnesium	Glucose 6-phosphataseHexokinaseDNA polymerase
Vanadium	vanabins
Manganese	Arginase
Iron	CatalaseHydrogenaseIRE-BPAconitase
Nickel	UreaseHydrogenase
Copper	Cytochrome oxidaseLaccase
Zinc	Alcohol dehydrogenase Carboxypeptidase Aminopeptidase Beta amyloid
Molybdenum	Nitrate reductase
Selenium	Glutathione peroxidase
various	MetallothioneinPhosphatase

Atmospheric Chemistry

Atmospheric chemistry is a branch of atmospheric science in which the chemistry of the Earth's atmosphere and that of other planets is studied. It is a multidisciplinary field of research and draws on environmental chemistry, physics, meteorology, computer modeling, oceanography, geology and volcanology and other disciplines. Research is increasingly connected with other areas of study such as climatology.

The composition and chemistry of the atmosphere is of importance for several reasons, but primarily because of the interactions between the atmosphere and living organisms. The composition of the Earth's atmosphere has been changed by human activity and some of these changes are harmful to human health, crops and ecosystems. Examples of problems which have been addressed by atmospheric chemistry include acid rain, ozone depletion, photochemical smog and global warming. Atmospheric chemistry seeks to understand the causes of these problems, and by obtaining a theoretical understanding of them, allow possible solutions to be tested and the effects of changes in government policy evaluated.

The ancient Greeks regarded air as one of the four elements, but the first scientific studies of atmospheric composition began in the 18th century. Chemists such as Joseph Priestley, Antoine Lavoisier and Henry Cavendish made the first measurements of the composition of the atmosphere.

In the late 19th and early 20th centuries interest shifted towards trace constituents with very small concentrations. One particularly important discovery for atmospheric chemistry was the discovery of ozone by Christian Friedrich Schoenbein in 1840. In the 20th century atmospheric science moved on from studying the composition of air to a consideration of how the concentrations of trace gases in the atmosphere have changed over time and the chemical processes which create and destroy compounds in the air. Two particularly important examples of this were the explanation of how the ozone layer is created and maintained by Sydney Chapman and Gordon Dobson, and the explanation of Photochemical smog by Arie Jan Haagen-Smit. Further studies on ozone issues led to the 1995 Nobel Prize in Chemistry award shared between Paul Crutzen, Mario Molina and Frank Sherwood Rowland.

In the 21st century the focus is now shifting again. Atmospheric chemistry is increasingly studied as one part of the Earth system. Instead of concentrating on atmospheric chemistry in isolation the focus is now on seeing it as one part of a single system with the rest of the atmosphere, biosphere and geosphere. An especially important driver for this is the links between chemistry and climate such as the effects of changing climate on the recovery of the ozone hole and vice versa but also interaction of the composition of the atmosphere with the oceans and terrestrial ecosystems.

Atmospheric Composition

Average composition of dry atmosphere, by volume

Gas	*Per NASA*
Nitrogen, N_2	78.084%
Oxygen, O_2	20.946%
Argon, Ar	0.934%
Minor constituents (in ppm)	
Carbon Dioxide, CO_2	383
Neon, Ne	18.18
Helium, He	5.24
Methane, CH_4	1.7
Krypton, Kr	1.14
Hydrogen, H_2	0.55
Water	
Water vapour	Highly variable;typically makes up about 1%

Notes: the concentration of CO_2 and CH_4 vary by season and location. The mean molecular mass of air is 28.97 g/mol.

Methodology

Observations, lab measurements and modeling are the three central elements in atmospheric chemistry. Progress in atmospheric chemistry is often driven by the interactions between these components and they form an integrated whole. For example observations may tell us that more of a chemical compound exists than previously thought possible. This will stimulate new modelling and laboratory studies which will increase our scientific understanding to a point where the observations can be explained.

Observation

Observations of atmospheric chemistry are essential to our understanding. Routine observations of chemical composition tell us about changes in atmospheric composition over time. One important example of this is the Keeling Curve- a series of measurements from 1958 to today which show a steady rise in of the concentration of carbon dioxide.

Observations of atmospheric chemistry are made in observatories such as that on Mauna Loa and on mobile platforms such as aircraft (e.g. the UK's Facility for Airborne Atmospheric Measurements), ships and balloons. Observations of atmospheric composition are increasingly made by satellites with important instruments such as GOME and MOPITT giving a global picture of air pollution and chemistry. Surface observations have the advantage that they provide long term records at high time resolution but are limited in the vertical and horizontal space they provide observations from. Some surface based instruments e.g. LIDAR can provide concentration profiles of chemical compounds and aerosol but are still restricted in the horizontal region they can cover. Many observations are available on line in Atmospheric Chemistry Observational Databases.

Lab Measurements

Measurements made in the laboratory are essential to our understanding of the sources and sinks of pollutants and naturally occurring compounds. Lab studies tell us which gases react with each other and how fast they react. Measurements of interest include reactions in the gas phase, on surfaces and in water. Also of high importance is photochemistry which quantifies how quickly molecules are split apart by sunlight and what the products are plus thermodynamic data such as Henry's law coefficients.

Modeling

In order to synthesise and test theoretical understanding of atmospheric chemistry, computer models (such as chemical transport models) are used. Numerical models solve the differential equations governing the concentrations of chemicals in the atmosphere. They can be very simple or very complicated. One common trade off in numerical models is between the number of chemical compounds and chemical reactions modelled versus the representation of transport and mixing in the atmosphere.

For example, a box model might include hundreds or even thousands of chemical reactions but will only have a very crude representation of mixing in the atmosphere. In contrast, 3D models represent many of the physical processes of the atmosphere but due to constraints on computer resources will have far fewer chemical reactions and compounds. Models can be used to interpret observations, test understanding of chemical reactions and predict future concentrations of chemical compounds in the atmosphere. One important current trend is for atmospheric chemistry modules to become one part of earth system models in which the links between climate, atmospheric composition and the biosphere can be studied.

Some models are constructed by automatic code generators (e.g. Autochem or KPP). In this approach a set of constituents are chosen and the automatic code generator will then select the reactions involving those constituents from a set of reaction databases. Once the reactions have been chosen the ordinary differential equations (ODE) that describe their time evolution can be automatically constructed.

Polyhedral Dynamics

A polyhedron (plural polyhedra or polyhedrons) is a geometric solid in three dimensions with flat faces and straight edges.

Defining a polyhedron as a solid bounded by flat faces and straight edges is not very precise and, to a modern mathematician, quite unsatisfactory. Grünbaum observed, "The Original Sin in the theory of polyhedra goes back to Euclid, and through Kepler, Poinsot, Cauchy and many others... [in that] at each stage... the writers failed to define what are the 'polyhedra'...." Since then rigorous definitions of "polyhedron" have been given within particular contexts. However such definitions are seldom compatible in other contexts.

Basis for Definition

Any polyhedron can be built up from different kinds of element or entity, each associated with a different number of dimensions:

- 3 dimensions: The body is bounded by the faces, and is usually the volume enclosed by them.

- 2 dimensions: A face is a *polygon* bounded by a circuit of edges, and usually including the flat (plane) region inside the boundary. These polygonal faces together make up the polyhedral surface.
- 1 dimension: An edge joins one vertex to another and one face to another, and is usually a line segment. The edges together make up the polyhedral skeleton.
- 0 dimensions: A vertex (plural vertices) is a corner point.
- -1 dimension: The null polytope is a kind of non-entity required by abstract theories.

More generally in mathematics and other disciplines, "polyhedron" is used to refer to a variety of related constructs, some geometric and others purely algebraic or abstract. A defining characteristic of almost all kinds of polyhedra is that just two faces join along any common edge. This ensures that the polyhedral surface is continuously connected and does not end abruptly or split off in different directions. A polyhedron is a 3-dimensional example of the more general polytope in any number of dimensions.

Characteristics

Names of Polyhedra

Polyhedra are often named according to the number of faces. The naming system is again based on Classical Greek, for example tetrahedron (4), pentahedron (5), hexahedron (6), heptahedron (7), triacontahedron (30), and so on. Often this is qualified by a description of the kinds of faces present, for example the Rhombic dodecahedron vs. the Pentagonal dodecahedron.

Other common names indicate that some operation has been performed on a simpler polyhedron, for example the truncated cube looks like a cube with its corners cut off, and has 14 faces (so it is also an example of a tetrakaidecahedron).

Some special polyhedra have grown their own names over the years, such as Miller's monster or the Szilassi polyhedron.

Edges

Edges have two important characteristics (unless the polyhedron is complex):

- An edge joins just two vertices.
- An edge joins just two faces.

These two characteristics are dual to each other.

Euler Characteristic

The Euler characteristic ÷ relates the number of vertices *V*, edges *E*, and faces *F* of a polyhedron:

For a simply connected polyhedron, ÷ = 2. For a detailed discussion, see Proofs and Refutations by Imre Lakatos.

Orientability

Some polyhedra, such as all convex polyhedra, have two distinct sides to their surface, for example one side can consistently be coloured black and the other white.

But for some polyhedra this is not possible, and the figure is said to be non-orientable. All polyhedra with odd-numbered Euler characteristic are non-orientable.

Vertex Figure

For every vertex one can define a vertex figure, which describes the local structure of the figure around the vertex. If the vertex figure is a regular polygon, then the vertex itself is said to be *regular*.

Duality

For every polyhedron there exists a dual polyhedron having:

- faces in place of the original's vertices and vice versa,
- the same number of edges
- the same Euler characteristic and orientability.

The dual of a convex polyhedron can be obtained by the process of polar reciprocation.

Traditional Polyhedra

In geometry, a polyhedron is traditionally a three-dimensional shape that is made up of a finite number of polygonal *faces* which are parts of planes; the faces meet in pairs along *edges* which are straight-line segments, and the edges meet in points called *vertices*. Cubes, prisms and pyramids are examples of polyhedra. The polyhedron surrounds a bounded volume in three-dimensional space; sometimes this interior volume is considered to be part of the polyhedron, sometimes only the surface is considered, and occasionally only the skeleton of edges. A polyhedron is said to be convex if its surface (comprising its faces, edges and vertices) does not intersect itself and the line segment joining any two points of the polyhedron is contained in the interior or surface.

Symmetrical Polyhedra

Many of the most studied polyhedra are highly symmetrical.

Of course it is easy to distort such polyhedra so they are no longer symmetrical. But where a polyhedral name is given, such as icosidodecahedron, the most symmetrical geometry is almost always implied, unless otherwise stated. Some of the most common names in particular are often used with "regular" in front or implied because for each there are different types which have little in common except for having the same number of faces.

Polyhedra of the highest symmetries have all of some kind of element-faces, edges and/or vertices, within a single symmetry orbit. There are various classes of such polyhedra:

- Isogonal or Vertex-transitive if all vertices are the same, in the sense that for any two vertices there exists a symmetry of the polyhedron mapping the first isometrically onto the second.
- Isotoxal or Edge-transitive if all edges are the same, in the sense that for any two edges there exists a symmetry of the polyhedron mapping the first isometrically onto the second.
- Isohedral or Face-transitive if all faces are the same, in the sense that for any two faces there exists a symmetry of the polyhedron mapping the first isometrically onto the second.
- Regular if it is vertex-transitive, edge-transitive and face-transitive (this implies that every face is the same regular polygon; it also implies that every vertex is regular).
- Quasi-regular if it is vertex-transitive and edge-transitive (and hence has regular faces) but not face-transitive. A quasi-regular dual is face-transitive and edge-transitive (and hence every vertex is regular) but not vertex-transitive.
- Semi-regular if it is vertex-transitive but not edge-transitive, and every face is a regular polygon. (This is one of several definitions of the term, depending on author. Some definitions overlap with the quasi-regular class). A semi-regular dual is face-transitive but not vertex-transitive, and every vertex is regular.
- Uniform if it is vertex-transitive and every face is a regular polygon, i.e. it is regular, quasi-regular or semi-regular. A uniform dual is face-transitive and has regular vertices, but is not necessarily vertex-transitive).
- Noble if it is face-transitive and vertex-transitive (but not necessarily edge-transitive). The regular polyhedra are also noble; they are the only noble uniform polyhedra.

A polyhedron can belong to the same overall symmetry group as one of higher symmetry, but will have several groups of elements (for example faces) in different symmetry orbits.

Uniform Polyhedra and their Duals

Uniform polyhedra are *vertex-transitive* and every face is a regular polygon. They may be regular, quasi-regular, or semi-regular, and may be convex or starry. The uniform duals are face-transitive and every vertex figure is a regular polygon.

Face-transitivity of a polyhedron corresponds to vertex-transitivity of the dual and conversely, and edge-transitivity of a polyhedron corresponds to edge-transitivity of the dual. In most duals of uniform polyhedra, faces are irregular polygons. The regular polyhedra are an exception, because they are dual to each other.

Each uniform polyhedron shares the same symmetry as its dual, with the symmetries of faces and vertices simply swapped over. Because of this some authorities regard the duals as uniform too. But this idea is not held widely: a polyhedron and its symmetries are not the same thing.

Noble Polyhedra

A noble polyhedron is both isohedral (equal-faced) and isogonal (equal-cornered). Besides the regular polyhedra, there are many other examples.

The dual of a noble polyhedron is also noble.

Symmetry Groups

The polyhedral symmetry groups (using Schoenflies notation) are all point groups and include:

- T- chiral tetrahedral symmetry; the rotation group for a regular tetrahedron; order 12.
- T_d- full tetrahedral symmetry; the symmetry group for a regular tetrahedron; order 24.
- T_h- pyritohedral symmetry; order 24. The symmetry of a pyritohedron.
- O- chiral octahedral symmetry;the rotation group of the cube and octahedron; order 24.
- O_h- full octahedral symmetry; the symmetry group of the cube and octahedron; order 48.
- I- chiral icosahedral symmetry; the rotation group of the icosahedron and the dodecahedron; order 60.

- I_h- full icosahedral symmetry; the symmetry group of the icosahedron and the dodecahedron; order 120.
- C_{nv}- *n*-fold pyramidal symmetry
- D_{nh}- *n*-fold prismatic symmetry
- D_{nv}- *n*-fold antiprismatic symmetry.

Those with chiral symmetry do not have reflection symmetry and hence have two enantiomorphous forms which are reflections of each other. The snub Archimedean polyhedra have this property.

Other Polyhedra with Regular Faces

Equal Regular Faces

A few families of polyhedra, where every face is the same kind of polygon:

- Deltahedra have equilateral triangles for faces.
- With regard to polyhedra whose faces are all squares: if coplanar faces are not allowed, even if they are disconnected, there is only the cube. Otherwise there is also the result of pasting six cubes to the sides of one, all seven of the same size; it has 30 square faces (counting disconnected faces in the same plane as separate). This can be extended in one, two, or three directions: we can consider the union of arbitrarily many copies of these structures, obtained by translations of (expressed in cube sizes) (2,0,0), (0,2,0), and/or (0,0,2), hence with each adjacent pair having one common cube. The result can be any connected set of cubes with positions (a,b,c), with integers a,b,c of which at most one is even.
- There is no special name for polyhedra whose faces are all equilateral pentagons or pentagrams. There are infinitely many of these, but only one is convex: the dodecahedron. The rest are assembled by (pasting) combinations of the regular polyhedra described earlier: the dodecahedron, the small stellated dodecahedron, the great stellated dodecahedron and the great icosahedron.

There exists no polyhedron whose faces are all identical and are regular polygons with six or more sides because the vertex of three regular hexagons defines a plane.

Deltahedra

A deltahedron (plural deltahedra) is a polyhedron whose faces are all equilateral triangles. There are infinitely many deltahedra, but only eight of these are convex:

- 3 regular convex polyhedra (3 of the Platonic solids)
    - o Tetrahedron
    - o Octahedron
    - o Icosahedron
- 5 non-uniform convex polyhedra (5 of the Johnson solids)
    - o Triangular dipyramid
    - o Pentagonal dipyramid
    - o Snub disphenoid
    - o Triaugmented triangular prism
    - o Gyroelongated square dipyramid.

Johnson Solids

Norman Johnson sought which convex non-uniform polyhedra had regular faces. In 1966, he published a list of 92 such solids, gave them names and numbers, and conjectured that there were no others. Victor Zalgaller proved in 1969 that the list of these Johnson solids was complete.

Other Important Families of Polyhedra

Pyramids

Pyramids include some of the most time-honoured and famous of all polyhedra.

Stellations and Facettings

Stellation of a polyhedron is the process of extending the faces (within their planes) so that they meet to form a new polyhedron.

It is the exact reciprocal to the process of facetting which is the process of removing parts of a polyhedron without creating any new vertices.

Zonohedra

A zonohedron is a convex polyhedron where every face is a polygon with inversion symmetry or, equivalently, symmetry under rotations through 180°.

Toroidal Polyhedra

A toroidal polyhedron is a polyhedron with an Euler characteristic of 0 or smaller, representing a torus surface.

Compounds

Polyhedral compounds are formed as compounds of two or more polyhedra.

These compounds often share the same vertices as other polyhedra and are often formed by stellation. Some are listed in the list of Wenninger polyhedron models.

Orthogonal Polyhedra

An orthogonal polyhedron is one all of whose faces meet at right angles, and all of whose edges are parallel to axes of a Cartesian coordinate system. Aside from a rectangular box, orthogonal polyhedra are nonconvex. They are the 3D analogs of 2D orthogonal polygons, also known as rectilinear polygons. Orthogonal polyhedra are used in computational geometry, where their constrained structure has enabled advances on problems unsolved for arbitrary polyhedra, for example, unfolding the surface of a polyhedron to a polygonal net.

Generalisations of Polyhedra

The name 'polyhedron' has come to be used for a variety of objects having similar structural properties to traditional polyhedra.

Apeirohedra

A classical polyhedral surface comprises finite, bounded plane regions, joined in pairs along edges. If such a surface extends indefinitely it is called an apeirohedron. Examples include:

- Tilings or tessellations of the plane.
- Sponge-like structures called infinite skew polyhedra.

Complex Polyhedra

A complex polyhedron is one which is constructed in complex Hilbert 3-space. This space has six dimensions: three real ones corresponding to ordinary space, with each accompanied by an imaginary dimension.

Curved Polyhedra

Some fields of study allow polyhedra to have curved faces and edges.

Spherical Polyhedra

The surface of a sphere may be divided by line segments into bounded regions, to form a spherical polyhedron. Much of the theory of symmetrical polyhedra is most conveniently derived in this way.

Spherical polyhedra have a long and respectable history:

- The first known man-made polyhedra are spherical polyhedra carved in stone.
- Poinsot used spherical polyhedra to discover the four regular star polyhedra.

- Coxeter used them to enumerate all but one of the uniform polyhedra.

Some polyhedra, such as hosohedra and dihedra, exist only as spherical polyhedra and have no flat-faced analogue.

Curved Spacefilling Polyhedra

Two important types are:

- Bubbles in froths and foams, such as Weaire-Phelan bubbles.
- Spacefilling forms used in architecture.

General Polyhedra

More recently mathematics has defined a polyhedron as a set in real affine (or Euclidean) space of any dimensional n that has flat sides. It may alternatively be defined as the union of a finite number of convex polyhedra, where a *convex polyhedron* is any set that is the intersection of a finite number of half-spaces.

It may be bounded or unbounded. In this meaning, a polytope is a bounded polyhedron. Analytically, such a convex polyhedron is expressed as the solution set for a system of linear inequalities. Defining polyhedra (and more generally polytopes) in this way provides a geometric perspective for problems in Linear programming.

Many traditional polyhedral forms are general polyhedra. Other examples include:

- A quadrant in the plane. For instance, the region of the cartesian plane consisting of all points above the horizontal axis and to the right of the vertical axis: $\{ (x, y) : x \geq 0, y \geq 0 \}$. Its sides are the two positive axes.
- An octant in Euclidean 3-space, $\{ (x, y, z) : x \geq 0, y \geq 0, z \geq 0\}$.
- A prism of infinite extent. For instance a doubly infinite square prism in 3-space, consisting of a square in the *xy*-plane swept along the *z*-axis: $\{ (x, y, z) : 0 \leq x \leq 1, 0 \leq y \leq 1 \}$.
- Each cell in a Voronoi tessellation is a convex polyhedron. In the Voronoi tessellation of a set S, the cell A corresponding to a point c"S is bounded (hence a traditional polyhedron) when c lies in the interior of the convex hull of S, and otherwise (when c lies on the boundary of the convex hull of S) A is unbounded.

Hollow Faced or Skeletal Polyhedra

It is not necessary to fill in the face of a figure before we can call it a polyhedron. For example Leonardo da Vinci devised frame models of the regular solids, which he drew for Pacioli's book *Divina Proportione*. In modern times, Branko Grünbaum (1994) made a special study of this class of polyhedra, in which he developed an early idea of abstract polyhedra. He defined a face as a cyclically ordered set of vertices, and allowed faces to be skew as well as planar.

Non-geometric Polyhedra

Various mathematical constructs have been found to have properties also present in traditional polyhedra.

Topological Polyhedra

A topological polytope is a topological space given along with a specific decomposition into shapes that are topologically equivalent to convex polytopes and that are attached to each other in a regular way.

Such a figure is called *simplicial* if each of its regions is a simplex, i.e. in an n-dimensional space each region has $n+1$ vertices. The dual of a simplicial polytope is called *simple*. Similarly, a widely studied class of polytopes (polyhedra) is that of cubical polyhedra, when the basic building block is an n-dimensional cube.

Abstract Polyhedra

An abstract polyhedron is a partially ordered set (poset) of elements whose partial ordering obeys certain rules. Theories differ in detail, but essentially the elements of the set correspond to the body, faces, edges and vertices of the polyhedron. The empty set corresponds to the null polytope, or *nullitope*, which has a dimensionality of –1.

These posets belong to the larger family of abstract polytopes in any number of dimensions.

Polyhedra as Graphs

Any polyhedron gives rise to a graph, or skeleton, with corresponding vertices and edges. Thus graph terminology and properties can be applied to polyhedra. For example:

- Due to Steinitz theorem convex polyhedra are in one-to-one correspondence with 3-connected planar graphs.
- The tetrahedron gives rise to a complete graph (K_4). It is the only polyhedron to do so.
- The octahedron gives rise to a strongly regular graph, because adjacent vertices always have two common neighbours, and non-adjacent vertices have four.

- The Archimedean solids give rise to regular graphs: 7 of the Archimedean solids are of degree 3, 4 of degree 4, and the remaining 2 are chiral pairs of degree 5.

History

Prehistory

Stones carved in shapes showing the symmetries of various polyhedra have been found in Scotland and may be as much a 4,000 years old. These stones show not only the form of various symmetrical polyehdra, but also the relations of duality amongst some of them (that is, that the centres of the faces of the cube gives the vertices of an octahedron, and so on).

Examples of these stones are on display in the John Evans room of the Ashmolean Museum at Oxford University. It is impossible to know why these objects were made, or how the sculptor gained the inspiration for them.

Other polyhedra have of course made their mark in architecture-cubes and cuboids being obvious examples, with the earliest four-sided pyramids of ancient Egypt also dating from the Stone Age.

The Etruscans preceded the Greeks in their awareness of at least some of the regular polyhedra, as evidenced by the discovery near Padua (in Northern Italy) in the late 19th century of a dodecahedron made of soapstone, and dating back more than 2,500 years (Lindemann, 1987). Pyritohedric crystals are found in northern Italy.

Greeks

The earliest known *written* records of these shapes come from Classical Greek authors, who also gave the first known mathematical description of them. The earlier Greeks were interested primarily in the convex regular polyhedra, which came to be known as the Platonic solids.

Pythagoras knew at least three of them, and Theaetetus (circa 417 B. C.) described all five. Eventually, Euclid described their construction in his *Elements*. Later, Archimedes expanded his study to the convex uniform polyhedra which now bear his name. His original work is lost and his solids come down to us through Pappus.

Chinese and Muslims

By 236 AD, in China Liu Hui was describing the dissection of the cube into its characteristic tetrahedron (orthoscheme) and related solids, using assemblages of these solids as the basis for calculating volumes of earth to be moved during engineering excavations.

After the end of the Classical era, Islamic scholars continued to take the Greek knowledge forward. The 9th century scholar Thabit ibn Qurra gave formulae for calculating the volumes of polyhedra such as truncated pyramids. Then in the 10th century Abu'l Wafa described the convex regular and quasiregular spherical polyhedra.

Renaissance

As with other areas of Greek thought maintained and enhanced by Islamic scholars, Western interest in polyhedra revived during the Italian Renaissance. Artists constructed skeletal polyhedra, depicting them from life as a part of their investigations into perspective. Several appear in marquetry panels of the period. Piero della Francesca gave the first written description of direct geometrical construction of such perspective views of polyhedra. Leonardo da Vinci made skeletal models of several polyhedra and drew illustrations of them for a book by Pacioli. A painting by an anonymous artist of Pacioli and a pupli depicts a glass rhombicuboctahedron half-filled with water.

As the Renaissance spread beyond Italy, later artists such as Wenzel Jamnitzer, Dürer and others also depicted polyhedra of various kinds, many of them novel, in imaginative etchings.

Star Polyhedra

For almost 2,000 years, the concept of a polyhedron as a convex solid had remained as developed by the ancient Greek mathematicians.

During the Renaissance star forms were discovered. A marble tarsia in the floor of St. Mark's Basilica, Venice, depicts a stellated dodecahedron. Artists such as Wenzel Jamnitzer delighted in depicting novel star-like forms of increasing complexity. Johannes Kepler realised that star polygons, typically pentagrams, could be used to build star polyhedra.

Some of these star polyhedra may have been discovered before Kepler's time, but he was the first to recognise that they could be considered "regular" if one removed the restriction that regular polytopes be convex.

Later, Louis Poinsot realised that star vertex figures (circuits around each corner) can also be used, and discovered the remaining two regular star polyhedra. Cauchy proved Poinsot's list complete, and Cayley gave them their accepted English names: (Kepler's) the small stellated dodecahedron and great stellated dodecahedron, and (Poinsot's) the great icosahedron and great dodecahedron. Collectively they are called the Kepler-Poinsot polyhedra.

The Kepler-Poinsot polyhedra may be constructed from the Platonic solids by a process called stellation. Most stellations are not regular. The study of stellations of the Platonic solids was given a big push by H. S. M. Coxeter and others in 1938, with the now famous paper *The 59 icosahedra*. This work has recently been re-published (Coxeter, 1999).

The reciprocal process to stellation is called facetting (or faceting). Every stellation of one polytope is dual, or reciprocal, to some facetting of the dual polytope. The regular star polyhedra can also be obtained by facetting the Platonic solids. Bridge 1974 listed the simpler facettings of the dodecahedron, and reciprocated them to discover a stellation of the icosahedron that was missing from the famous "59". More have been discovered since, and the story is not yet ended.

Polyhedral Compound

A polyhedral compound is a polyhedron that is itself composed of several other polyhedra sharing a common centre. They are the three-dimensional analogs of polygonal compounds such as the hexagram.

Neighbouring vertices of a compound can be connected to form a convex polyhedron called the convex hull. The compound is a facetting of the convex hull. Another convex polyhedron is formed by the small central space common to all members of the compound. This polyhedron can be considered the core for a set of stellations including this compound.

Dithiocarbamate Complexes

A dithiocarbamate is a functional group in organic chemistry. It is the analog of a carbamate in which both oxygen atoms are replaced by sulfur atoms. Sodium diethyldithiocarbamate is a common ligand in inorganic chemistry. Most primary and secondary amines react with carbon disulfide and sodium hydroxide to form dithiocarbamates. They are used as ligands for chelating metals. Dithiocarbamates readily react with many metal salts such as copper, ferrous, ferric, cobaltous, and nickel salts and are mostly found as octahedral complexes.

Sodium Diethyldithiocarbamate

Sodium diethyldithiocarbamate is the organosulfur compound with the formula $NaS_2CN(C_2H_5)_2$.

Preparation

This salt is obtained by treating carbon disulfide with diethylamine in the presence of sodium hydroxide:

$$CS_2 + HN(C_2H_5)_2 + NaOH \rightarrow NaS_2CN(C_2H_5)_2 + H_2O$$

Other dithiocarbamates can be prepared similarly from secondary amines and carbon disulfide. They are used as chelating agents for transition metal ions and as precursors to herbicides and vulcanization reagents.

Oxidation to Thiuram Disulfide

Oxidation of sodium diethyldithiocarbamate gives the disulfide, also called a thiuram disulfide (Et = ethyl):

$$2\ NaS_2CNEt_2 + I_2 \rightarrow Et_2NC(S)S\text{-}SC(S)NEt_2 + 2\ NaI$$

This disulfide is marketed as an anti-alcoholism drug under the labels Antabuse and Disulfiram. Chlorination of the above-mentioned thiuram disulfide affords the thiocarbamoyl chloride.

Ligand Bonding

The diethyldithiocarbamate ion chelates to many "softer" metals via the two sulfur atoms. Other more complicated bonding modes are known including binding as unidentate ligand and a bridging ligand using one or both sulfur atoms.

Spin Trapping of Nitric Oxide Radicals

Complexes of Dithiocarbamates with iron provide one of the very few methods to study the formation of nitric oxide (NO) radicals in biological materials. Although the lifetime of NO in tissues is too short to allow detection of this radical itself, NO readily binds to iron-dithiocarbamate complexes. The resulting mono-nitrosyl-iron complex (MNIC) is stable, and may be detected with Electron Paramagnetic Resonance (EPR) spectroscopy.

In Cancer

The effect of diethyldithiocarbamate of chelating zinc inhibits metalloproteinases, which in turn prevents the degradation of extracellular matrix, which is an initial step in cancer metastasis and angiogenesis.

Antioxidant

Diethyldithiocarbamate inhibits superoxide dismutase, which can both have antioxidant and oxidant effects on cells, depending on the time of administration.

Main group dithiocarbamate complexes find wide-ranging applications in materials and separation science, and have potential use as chemotherapeutics, pesticides, and fungicides. The literature on main group dithiocarbamates as a whole has not been reviewed extensively since the 1970s (1, 2) despite the large number of publications that have appeared subsequently. From an inorganic chemistry stand point, dithiocarbamates are highly versatile ligands toward maigroup metals.

They can stabilize a variety of oxidation states and coordination geometries, and seemingly small modifications to the ligand can lead to significant changes in the structure–behaviour of the complexes formed. This chapter focuses primarily on structural aspect of main group dithiocarbamate complexes, covering the essential literature from 1978 to 2003. For the purposes of this chapter, the zinc triad of elements is not considered as being main group:

The structural parameters of the dithiocarbamate ligands themselves are not modified significantly on coordination to main group elements. Distances (A °) and angles (_) are in the range: C_N(R2) ¼ 1.24–1.52 (1.33 mean); C_S ¼ 1.52–1.82 (1.72 mean); SCS ¼ 110.1–128.9 (118.6 mean). The two C_S distances are often slightly different, indicating some charge localization: as one would expect, the shorter C_S are generally associated with the S atom that is least strongly associated with the metal centre. The SCS bond angle generally increases in line with the size of the metal to which the dithiocarbamate is coordinated.

Extensive use has also been made of infrared (IR) spectroscopy for the characterization of dithiocarbamate complexes. Particular diagnostic use has been made of the C_N(R2) and C__S stretching modes, which fall typically in the range 1500 _ 50 and 980 _ 50 cm_1, respectively. The occurrence of more than one C__S stretching band has been used to imply a monodentate or highly anisobidentate bonding mode for the dithiocarbamate ligand; however, this is not unambiguous and should not be relied upon.

s-Block Metals

The dithiocarbamates of the s-block elements, prepared by reaction of the appropriate amine and CS2 in the presence of the metal cation, are water soluble, ionic compounds. The structures of a number of hydrated compounds have been determined by X-ray crystallography (3–26). Although there are some exceptions, except for the heavier metal ions, there is generally little or no direct interaction between metal ion and the essentially planar dithiocarbamate anion in the solid state. Structural data show that the C_S bond distances of the free dithiocarbamate moieties lie in the range 1.65–1.81A °; one C_S bond is often significantly shorter than the other, indicating some localization of the negative charge. The C_N distances and SCS angles lie in the range 1.32–1.43A ° and 117.0–128.9_, respectively. The SCS bond angle generally decreases with increasing steric bulk of the N substituents, as a result of intermolecular S...H_C interactions.

Under anhydrous conditions, dithiocarbamates can coordinate to lightes-block metal ions. Insertion of CS2 into the Be_N bonds of bis(diisopropylamino) beryllium yields the corresponding bis(diisopropyldithiocarbamate)beryllium compound.

The four S atoms of the two dithiocarbamate ligands coordinate to Be, giving a distorted tetrahedral geometry at the metal ion. The 9Be nuclear magnetic resonance (NMR) spectrum of [Be(S2CNi-Pr2)2] in cyclohexane displays a single resonance at d 1.55, typical of four-coordinate Be, indicating that the solid-state structure is retained in solution. Similarly, lithiation of diphenylamine followed by reaction with CS2 in tetrahydrofuran (THF) gives Ph2NCS2Li_2THF, in which the diphenyldithiocarbamate anion chelates the Liþ ion.

Protein-surface Interactions

The study of protein–surface interactions represents one of the most important topics in the field of biomaterialsand as such, it has been a focus of intensive study for several decades. The reason for the great interest in this topic is the realization that protein–surface interactions are fundamentally responsible for the biocompatibility of medical devices, or the lack thereof.

When a solid material (e.g., a catheter, stent, hip joint replacement, or tissue engineering substrate) comes in contact with a fluid that contains soluble proteins (e.g., blood, interstitial fluid, cell culture media), proteins rapidly adsorb onto the surface of the material, saturating the surface within a time frame of seconds to minutes.

Therefore, when living cells (which are much larger than proteins and thus much morslowly moving) approach the biomaterial surface, they do not actually contact the molecular structure of the material surface itself, but rather they contact and interact with the molecular structure of the adsorbed protein layer.

Cells, of course, cannot "see" the adsorbed protein layer, but rather they interrogate their surroundings by way of membrane-bound receptors that can bind to specific bioactive features presented by the adsorbed proteins.

Then, through a series of orchestrated molecular mechanisms, these receptor–protein binding events are transduced through the cell membrane in a manner that stimulates specific intracellular processes that then determine a cell's response. Accordingly, at the most fundamental level the key to controlling cellular response is to control the type of bioactive sites that are presented by the adsorbed layer of proteins.

This, in turn, can be controlled by controlling the amounts and the types of proteins that are adsorbed and their orientation, conformation, and packing arrangement on the biomaterial surface.

While this is conceptually simple to understand, the numerous types of soluble proteins contained in physiological fluids combined with their structural complexity has made, and continues to make, this an extremely challenging problem.

Many excellent reviews and entire books have been written over the past couple of decades that address the multitude of issues related to the interaction of proteins with surfaces. Several of these reviews and books are referenced in the "Further Readings" section at the end of this article and the readers interested in this topic are encouraged to seek out these additional sources of information. Given the vast number of studies that have been conducted on this topic, it is of course impossible to provide a complete review of the literature in one article. Instead, the goal here is to provide an overview of the basic understanding that has been achieved over the years regarding how proteins interact with surfaces.

This will be accomplished by first providing an overview of protein adsorption processes and then by addressing the special case of the design of surfaces to prevent protein adsorption.

Following these topics, attention will be focused on highlighting several of the most interesting relatively recent techniques that have been developed and applied to further our understanding of the submolecular-level mechanisms involved in how surface chemistry influences the orientation, conformation, and organization of adsorbed proteins.

The continued development of our understanding of these processes is critical if we are to get beyond the current era of surface design largely by trial and error, and move into an era where surfaces are proactively designed to directly control adsorbed protein bioactivity, and thereby control cellular response.

This article will then close with a conclusion section that addresses future directions toward the continuing goal of the development of truly biocompatible materials for the design of medical devices for improved patient care.

Overview of Protein–Surface Interactions

As introduced in the prior section, the surfaces of synthetic biomaterials (e.g., polymers, metals, and ceramics) are generally not bioactive themselves. Rather, surface bioactivity is provided by the proteins that adsorb to the biomaterial surface following exposure of the surface to biological fluids.

The types and the amounts of proteins adsorbed determine the types and surface density of the bioactive sites that may be available for cell interactions, and the orientation, conformation, and packing density of the adsorbed proteins determine whether the available bioactive sites are presented in a manner such that they can be recognized by the membrane-bound receptors of cells as they interrogate the adsorbed protein layer.

Accordingly, to control cellular response, it is important to first understand how surface chemistry and surface topology influence the formation of the adsorbed protein layer and the bioactive sites presented by this layer. Although the understanding of protein adsorption to biomaterial surfaces is still far from complete, research over the past several decades has led to a general understanding of the complex and multifaceted processes involved in the interactions between proteins and surfaces. In this section, an overview of protein structure is first presented to establish a framework from which protein adsorption behaviour can begin to be understood.

Following this, an overview of the current understanding of protein adsorption behaviour to biomaterial surfaces is presented, initially for the relatively simple case involving protein adsorption from single-component protein solutions, and then for the more complex situation involving competitive adsorption from multicomponent protein solutions. This section will then be completed by addressing the special case of the development of nonfouling surfaces that are highly resistant to protein adsorption.

Protein Adsorption Behaviour from Single-Component Protein Solutions

Where P is the protein concentration in solution, S is the concentration of available surface area sites for adsorption, P _ S is the concentration of reversibly adsorbed protein on the surface with kf and kr designating reaction rate constants for the forward and reverse reactions, respectively, and P _ S is the concentration of irreversibly adsorbed protein on the surface with ki designating the reaction rate constant for the transition of the adsorbed protein from a reversibly bound to an irreversibly bound state.

Water is shown complexed to the protein and surface, which is then released to bulk solution during the various stages of adsorption. This is included to emphasize its importance as a molecular constituent in the adsorption process rather than simply being an inert media that the protein and surface happen to be immersed in.

Because of the presence of the irreversible step shown at the right-hand side of this equation, a surface that adsorbs a protein in a manner that includes this part of the process will inevitably become saturated with the adsorbed protein if sufficient interaction time is provided.

The state of the final adsorbed protein layer will be determined by the manner in which the proteins are able solid material surface.

Matrix Characterization

Dioxygen Complex

Dioxygen complexes are coordination compounds that contain O_2 as a ligand. The study of these compounds is inspired by oxygen-carrying proteins such as myoglobin, hemoglobin, hemerythrin, and hemocyanin. Several transition metals form complexes with O_2, and many of these complexes form reversibly. The binding of O_2 is the first step in many important phenomena, such as cellular respiration, corrosion, and industrial chemistry. The first synthetic oxygen complex was demonstrated in 1938 with cobalt(II) complex reversibly bound O_2.

Mononuclear Complexes of O_2

O_2 binds to a single metal centre either "end-on" (η^1-) or "side-on" (η^2-). The bonding and structures of these compounds are usually evaluated by single-crystal X-ray crystallography, focusing both on the overall geometry as well as the O—O distances, which reveals its bond order.

Complexes of η^1-O_2 Ligands

O_2 adducts derived from cobalt(II) and iron(II) porphyrin complexes and related anionic ligands exhibit this bonding mode. Myoglobin and hemoglobin are famous examples, and many synthetic analogues have been described that behave similarly. Binding of O_2 is usually described as proceeding via electron transfer from the metal(II) centre to give superoxide (O_2^-) complexes of metal(III) centres.

Complexes of η^2-O_2 Ligands

η^2- bonding is the most common motif seen in coordination chemistry of dioxygen. Such complexes can generated by treating low-valent metal complexes with gaseous oxygen. For example, Vaska's complex reversibly binds O_2 (Ph = C_6H_5):

$$IrCl(CO)(PPh_3)_2 + O_2 \ IrCl(CO)(PPh_3)_2O_2$$

The conversion is described as a 2 e^- redox process: Ir(I) converts to Ir(III) as dioxygen converts to peroxide. Since O_2 has a triplet ground state and Vaska's complex is a singlet, the reaction is slower than when singlet oxygen is used.

Complexes containing η^2-O_2 ligands are fairly common, but most are generated using hydrogen peroxide, not O_2. Chromate ($[CrO_4]^{2-}$) can for example be converted to the tetraperoxide $[Cr(O_2)_4]^{2-}$. The reaction of hydrogen peroxide with aqueous titanium(IV) gives a brightly coloured peroxy complex that is a useful test for titanium as well as hydrogen peroxide.

Binuclear Complexes of O_2

O_2 can bind to one metal of a bimetallic unit via the same modes discussed above for mononuclear complexes. A well known example in nature is hemerythrin, which features a diiron carboxylate that binds O_2 at one Fe centre.

Dinuclear complexes can also cooperate in the binding, although the initial attack of O_2 probably occurs at a single metal. These binding modes include μ_2-η^2,η^2-, μ_2-η^1,η^1-, and μ_2-η^1,η^2-. Depending on the degree of electron-transfer from the dimetal unit, these O_2 ligands can again be described as peroxo or superoxo. In nature, such dinuclear dioxygen complexes often feature copper.

Relationship to other Oxygenic Ligands and Applications

Dioxygen complexes are the precursors to other families of oxygenic ligands. Metal oxo compounds arise from the cleavage of the O-O bond after complexation. Hydroperoxo complexes are generated in the course of the reduction of dioxygen by metals. The reduction of O_2 by metal catalysts is a key half-reaction in fuel cells.

Metal-catalysed oxidations with O_2 proceed via the intermediacy of dioxygen complexes, although the actual oxidants are often oxo derivatives. The reversible binding of O_2 to metal complexes has been used as a means to purify oxygen from air, but cryogenic distillation of liquid air remains the dominant technology.

Coordination Complex

In chemistry, a coordination complex or metal complex, is a structure consisting of a central atom or ion (usually metallic), bonded to a surrounding array of molecules or anions (ligands, complexing agents). The atom within a ligand that is directly bonded to the central atom or ion is called the donor atom. Polydentate (multiple bonded) ligands can form a chelate complex. A ligand donates at least one pair of electrons to the central atom/ion.

Compounds that contain a coordination complex are called coordination compounds. The central atom or ion, together with all ligands form the coordination sphere.

Coordination refers to the "coordinate covalent bonds" (dipolar bonds) between the ligands and the central atom. Originally, a complex implied a reversible association of molecules, atoms, or ions through such weak chemical bonds. As applied to coordination chemistry, this meaning has evolved. Some metal complexes are formed virtually irreversibly and many are bound together by bonds that are quite strong.

History

Coordination complexes were known-although not understood in any sense-since the beginning of chemistry, e.g. Prussian blue and copper vitriol. The key breakthrough occurred when Alfred Werner proposed in 1893 that Co(III) bears six ligands in an octahedral geometry. His theory allows one to understand the difference between coordinated and ionic in a compound, for example chloride in the cobalt ammine chlorides and to explain many of the previously inexplicable isomers.

In 1914, Werner resolved the first coordination complex, called hexol, into optical isomers, overthrowing the theory that chirality was necessarily associated with carbon compounds.

Structure of Coordination Complexes

The ions or molecules surrounding the central atom are called ligands. Ligands are generally bound to the central atom by a coordinate covalent bond (donating electrons from a lone electron pair into an empty metal orbital), and are said to be coordinated to the atom. There are also organic ligands such as alkenes whose pi bonds can coordinate to empty metal orbitals. An example is ethene in the complex known as Zeise's salt, $K^+[PtCl_3(C_2H_4)]^-$.

Geometry

In coordination chemistry, a structure is first described by its coordination number, the number of ligands attached to the metal (more specifically, the number of σ-type bonds between ligand(s) and the central atom).

Usually one can count the ligands attached, but sometimes even the counting can become ambiguous. Coordination numbers are normally between two and nine, but large numbers of ligands are not uncommon for the lanthanides and actinides. The number of bonds depends on the size, charge, and electron configuration of the metal ion and the ligands. Metal ions may have more than one coordination number.

Typically the chemistry of complexes is dominated by interactions between s and p molecular orbitals of the ligands and the d orbitals of the metal ions. The s, p, and d orbitals of the metal can accommodate 18 electrons.

The maximum coordination number for a certain metal is thus related to the electronic configuration of the metal ion (more specifically, the number of empty orbitals) and to the ratio of the size of the ligands and the metal ion. Large metals and small ligands lead to high coordination numbers, e.g. $[Mo(CN)_8]^{4-}$. Small metals with large ligands lead to low coordination numbers, e.g. $Pt[P(CMe_3)]_2$. Due to their large size, lanthanides, actinides, and early transition metals tend to have high coordination numbers.

Different ligand structural arrangements result from the coordination number. Most structures follow the points-on-a-sphere pattern (or, as if the central atom were in the middle of a polyhedron where the corners of that shape are the locations of the ligands), where orbital overlap (between ligand and metal orbitals) and ligand-ligand repulsions tend to lead to certain regular geometries.

The most observed geometries are listed below, but there are many cases which deviate from a regular geometry, e.g. due to the use of ligands of different types (which results in irregular bond lengths; the coordination atoms do not follow a points-on-a-sphere pattern), due to the size of ligands, or due to electronic effects:

- Linear for two-coordination,
- Trigonal planar for three-coordination,

- Tetrahedral or square planar for four-coordination
- Trigonal bipyramidal or square pyramidal for five-coordination,
- Octahedral (orthogonal) or trigonal prismatic for six-coordination,
- Pentagonal bipyramidal for seven-coordination,
- Square antiprismatic for eight-coordination, and
- Tri-capped trigonal prismatic (Triaugmented triangular prism) for nine coordination.

Some exceptions and provisions should be noted:

- The idealized descriptions of 5-, 7-, 8-, and 9- coordination are often indistinct geometrically from alternative structures with slightly different L-M-L (ligand-metal-ligand) angles. The classic example of this is the difference between square pyramidal and trigonal bipyramidal structures.
- Due to special electronic effects such as (second-order) Jahn-Teller stabilization, certain geometries are stabilized relative to the other possibilities, e.g. for some compounds the trigonal prismatic geometry is stabilized relative to octahedral structures for six-coordination.

Isomerism

The arrangement of the ligands is fixed for a given complex, but in some cases it is mutable by a reaction that forms another stable isomer.

There exist many kinds of isomerism in coordination complexes, just as in many other compounds.

Stereoisomerism

Stereoisomerism occurs with the same bonds in different orientations relative to one another. Stereoisomerism can be further classified into:

Cis-trans Isomerism and Facial-meridional Isomerism

Cis-trans isomerism occurs in octahedral and square planar complexes (but not tetrahedral). When two ligands are mutually adjacent they are said to be cis, when opposite each other, trans. When three identical ligands occupy one face of an octahedron, the isomer is said to be facial, or fac.

In a *fac* isomer, any two identical ligands are adjacent or *cis* to each other. If these three ligands and the metal ion are in one plane, the isomer is said to be meridional, or mer. A *mer* isomer can be considered as a combination of a *trans* and a *cis*, since it contains both trans and cis pairs of identical ligands.

Optical Isomerism

Optical isomerism occurs when the mirror image of a compound is not superimposable with the original compound. It is so called because such isomers are optically active, that is, they rotate the plane of polarized light.

The symbol Λ (*lambda*) is used as a prefix to describe the left-handed propeller twist formed by three bidentate ligands, as shown. Similarly, the symbol □Δ (*delta*) is used as a prefix for the right-handed propeller twist.

Structural Isomerism

Structural isomerism occurs when the bonds are themselves different. Linkage isomerism is only one of several types of structural isomerism in coordination complexes (as well as other classes of chemical compounds).

Linkage isomerism occurs with ambidentate ligands which can bind in more than one place. For example, NO_2 is an ambidentate ligand: it can bind to a metal at either the N atom or at an O atom.

Chapter 2

Microbial Metabolism

Microbial metabolism is the means by which a microbe obtains the energy and nutrients (e.g. carbon) it needs to live and reproduce. Microbes use many different types of metabolic strategies and species can often be differentiated from each other based on metabolic characteristics. The specific metabolic properties of a microbe are the major factors in determining that microbe's ecological niche, and often allow for that microbe to be useful in industrial processes or responsible for biogeochemical cycles.

Types of Microbial Metabolism

All microbial metabolisms can be arranged according to three principles:

How the organism obtains carbon for synthesising cell mass:

Autotroph

An autotroph, also called a producer, is an organism that produces complex organic compounds (such as carbohydrates, fats, and proteins) from simple inorganic molecules using energy from light (by photosynthesis) or inorganic chemical reactions (chemosynthesis). They are the producers in a food chain, such as plants on land or algae in water. They are able to make their own food and can fix carbon. Therefore, they do not utilize organic compounds as an energy source or a carbon source. Autotrophs can reduce carbon dioxide (add hydrogen to it) to make organic compounds. The reduction of carbon dioxide, a low-energy compound, creates a store of chemical energy. Most autotrophs use water as the reducing agent, but some can use other hydrogen compounds such as hydrogen sulfide. An autotroph converts physical energy from sun light (in case of green plants) into chemical energy in the form of reduced carbon. Autotroph can be phototrophs or lithotrophs (chemoautotrophs). Phototrophs use light as an energy source, while lithotrophs oxidize inorganic compounds, such as hydrogen sulfide, elemental sulfur, ammonium and ferrous iron. Phototrophs and lithotrophs use a portion of the ATP produced during photosynthesis or the oxidation of inorganic compounds to reduce $NADP^+$ to NADPH in order to form organic compounds.

Ecology

Autotrophs are fundamental to the food chains of all ecosystems in the world. They take energy from the environment in the form of sunlight or inorganic chemicals and use it to create energy-rich molecules such as carbohydrates. This mechanism is called primary production. Other organisms, called heterotrophs, take in autotrophs as food to carry out functions necessary for their life. Thus, heterotrophs — all animals, almost all fungi, as well as most bacteria and protozoa — depend on autotrophs for the energy and raw materials they need. Heterotrophs obtain energy by breaking down organic molecules (carbohydrates, fats, and proteins) obtained in food. Carnivorous organisms ultimately rely on autotrophs because the nutrients obtained from their heterotroph prey come from autotrophs they consumed.

Variants

Some organisms rely on organic compounds as a source of carbon, but are able to use light or inorganic compounds as a source of energy. Such organisms are not defined as autotrophic, but rather as heterotrophic. An organism that obtains carbon from organic compounds but obtains energy from light is called a *photoheterotroph*, while an organism that obtains carbon from organic compounds but obtains energy from the oxidation of inorganic compounds is termed a *chemoheterotroph* or *chemolithoheterotroph.*

Evidence suggests that some fungi may also obtain energy from radiation. Such radiotrophic fungi were found growing inside a reactor of the Chernobyl nuclear power plant.

Heterotroph

A heterotroph is an organism that cannot fix carbon and uses organic carbon for growth. This contrasts with autotrophs, such as plants and algae, which can use energy from sunlight (photoautotrophs) or inorganic compounds (lithoautotrophs) to produce organic compounds such as carbohydrates, fats, and proteins from inorganic carbon dioxide.

These reduced carbon compounds can be used as an energy source by the autotroph and provide the energy in food consumed by heterotrophs. Ninety-five percent or more of all types of living organisms are heterotrophic.

Types

Heterotrophs must utilize organic compounds as a source of carbon, but can obtain energy either from light or from chemical compounds. They can be divided into two broad classes: photoheterotrophs and chemoheterotrophs. Photoheterotrophs, including most purple bacteria and green bacteria, produce ATP from light and use organic compounds to build structures.

They consume little or none of the energy produced during photosynthesis to reduce $NADP^+$ to NADPH for use in the Calvin cycle, as they do not need to use the Calvin cycle if carbohydrates are available in their diets. Chemoheterotrophs produce ATP by oxidizing chemical substances. There are two types of chemoheterotrophs: chemoorganoheterotrophs and chemolithoheterotrophs.

Chemoorganoheterotrophs (or simply organotrophs) exploit reduced carbon compounds as energy sources, such as carbohydrates, fats, and proteins from plants and animals. Chemolithoheterotrophs (or lithotrophic heterotrophs) such as colorless sulfur bacteria (e.g., *Beggiatoa* and *Thiobacillus*) and sulfate-reducing bacteria utilize inorganic substances to produce ATP, including hydrogen sulfide, elemental sulfur, thiosulfate, and molecular hydrogen. They use organic compounds to build structures. They do not fix carbon dioxide and apparently do not have the Calvin cycle. Chemolithoheterotrophs can be distinguished from mixotrophs (or facultative chemolithotroph), which can utilize either carbon dioxide or organic carbon as the carbon source.

Heterotrophs, by consuming reduced carbon compounds, are able to use all the energy that they obtain from food for growth and reproduction, unlike autotrophs, which must use some of their energy for carbon fixation. Heterotrophs are unable to make their own food, however, and whether using organic or inorganic energy sources, they can die from a lack of food. This applies not only to animals and fungi but also to bacteria.

Most heterotrophs are chemoorganoheterotrophs (or simply organotrophs) and utilize organic compounds both as a carbon source and an energy source. The term "heterotroph" very often refers to chemoorganoheterotrophs. Heterotrophs function as consumers in food chains: they obtain organic carbon by eating other heterotrophs or autotrophs. They break down complex organic compounds (e.g., carbohydrates, fats, and proteins) produced by autotrophs into simpler compounds (e.g., carbohydrates into glucose, fats into fatty acids and glycerol, and proteins into amino acids). They release energy by oxidizing carbon and hydrogen atoms present in carbohydrates, lipids, and proteins to carbon dioxide and water, respectively.

All animals and fungi are heterotrophic, as well as most protists and prokaryotes. Some animals, such as corals, form symbiotic relationships with autotrophs and obtain organic carbon in this way. Furthermore, some parasitic plants have also turned fully or partially heterotrophic, while carnivorous plants consume animals to augment their nitrogen supply while remaining autotrophic.

Mixotroph

A mixotroph is a microorganism that can use a mix of different sources of energy and carbon. Possible are alternations between photo- and chemotrophy, between litho- and organotrophy, between auto- and heterotrophy or a combination of it. Mixotrophs can be either eukaryotic or prokaryotic. They can take advantage of different environmental conditions. If a trophic mode is obligate, then it is always necessary for sustaining growth and maintenance; if facultative, it can be used as a supplemental source. Some organisms have incomplete Calvin cycles, so they are incapable of fixing carbon dioxide and must use organic carbon sources.

Examples

- *Paracoccus pantotrophus* is a bacterium that can live as chemoorganoheterotroph, whereby a large variety of organic compounds can be metabolized. Also a facultative chemolithoautotrophic metabolism is possible, as seen in colorless sulfur bacteria (some *Thiobacillus*), whereby sulfur compounds such as hydrogen sulfide, elemental sulfur, or thiosulfate are oxidized to sulfate. The sulfur compounds serve as electron donors and are consumed to produce ATP. The carbon source for these organisms can be carbon dioxide (autotrophy) or organic carbon (heterotrophy).

Organoheterotrophy can occur under aerobic or under anaerobic conditions; lithoautotrophy takes place aerobically. How the organism obtains reducing equivalents used either in energy conservation or in biosynthetic reactions.

Lithotroph

A lithotroph is an organism that uses an inorganic substrate (usually of mineral origin) to obtain reducing equivalents for use in biosynthesis (e.g., carbon dioxide fixation) or energy conservation via aerobic or anaerobic respiration. Known lithotrophs are exclusively microbes; No known macrofauna possesses the ability to utilize inorganic compounds as energy sources. Macrofauna and lithotrophs can form symbiotic relationships, in which case the lithotrophs are called "prokaryotic symbionts." An example of this is chemolithotrophic bacteria in deep sea worms or plastids, which are organelles within plant cells that may have evolved from photolithotrophic cyanobacteria-like organisms. Lithotrophs belong to either the domain Bacteria or the domain Archaea. The term "Lithotroph" is created from the terms 'lithos' (rock) and 'troph' (consumer), meaning the "eaters of rock." Many lithoautotrophs are extremophiles, but this is not universally so.

The opposite of lithotroph is organotroph - an organism which gets its energy from the catabolism of organic compounds.

Biochemistry

Lithotrophs consume reduced compounds (rich in electrons).

Chemolithotrophs

In chemolithotrophs, the compounds - the electron donors - are oxidized in the cell, and the electrons are channeled into respiratory chains, ultimately producing ATP. The electron acceptor can be oxygen (in aerobic bacteria), but a variety of other electron acceptors, organic and inorganic, are also used by various species.

Photolithotrophs

Photolithotrophs obtain energy from light and therefore use inorganic electron donors only to fuel biosynthetic reactions (e.g., carbon dioxide fixation in lithoautotrophs).

Lithoheterotrophs Versus Lithoautotrophs

Lithotrophic bacteria cannot use, of course, their inorganic energy source as a carbon source for the synthesis of their cells. They choose one of three options:

- Lithoheterotrophs do not have the possibility to fix carbon dioxide and must consume additional organic compounds in order to break them apart and use their carbon. Only few bacteria are fully heterolithotrophic.

- Lithoautotrophs are able to use carbon dioxide from the air as carbon source, the same way plants do.
- Mixotrophs will take up and utilise organic material to complement their carbon dioxide fixation source (mix between autotrophy and heterotrophy). Many lithotrophs are recognised as mixotrophic in regard of their C-metabolism.

Chemolithotrophs Versus Photolithotrophs

In addition to this division, lithotrophs differ in the initial energy source which initiates ATP production:

- Chemolithotrophs use the above-mentioned inorganic compounds for aerobic or anaerobic respiration. The energy produced by the oxidation of these compounds is enough for ATP production. Some of the electrons derived from the inorganic donors also need to be channeled into biosynthesis. Mostly, additional energy has to be invested to transform these reducing equivalents to the forms and redox potentials needed (mostly NADH or NADPH), which occurs by reverse electron transfer reactions.
- Photolithotrophs use light as energy source. These bacteria are photosynthetic; photolithotrophic bacteria are found in the purple bacteria (e.g., Chromatiaceae), green bacteria (Chlorobiaceae and Chloroflexi) and Cyanobacteria. The electrons obtained from the electron donors (purple and green bacteria oxidize sulfide, sulfur, sulfite, iron or hydrogen; cyanobacteria extract reducing equivalents from water, i.e., oxidise water to oxygen) are not used for ATP production (as long as there is light); they are used in biosynthetic reactions. Some photolithotrophs shift over to chemolithotrophic metabolism in the dark.

Geological Significance

Lithotrophs participate in many geological processes, such as the weathering of parent material (bedrock) to form soil, as well as biogeochemical cycling of sulfur, nitrogen, and other elements. They may be present in the deep terrestrial subsurface (they have been found well over 3 km below the surface of the planet), in soils, and in endolith communities. As they are responsible for the liberation of many crucial nutrients, and participate in the formation of soil, lithotrophs play a critical role in the maintenance of life on Earth.

Lithotrophic microbial consortia are responsible for the phenomenon known as acid mine drainage, whereby energy-rich pyrites and other reduced sulfur compounds present in mine tailing heaps and in exposed rock faces is metabolized to form sulfates, thereby forming potentially toxic sulfuric acid. Acid mine drainage drastically alters the acidity and chemistry of groundwater and streams, and may endanger plant and animal populations. Activities similar to acid mine drainage, but on a much lower scale, are also found in natural conditions such as the rocky beds of glaciers, in soil and talus, on stone monuments and buildings and in the deep subsurface.

Organotroph

An organotroph is an organism that obtains hydrogen or electrons from organic substrates (a form of a chemotroph).

Antonym: Lithotroph

How the organism obtains energy for living and growing.

Chemotroph

Chemotrophs are organisms that obtain energy by the oxidation of electron donors in their environments. These molecules can be organic (chemoorganotrophs) or inorganic (chemolithotrophs). The chemotroph designation is in contrast to phototrophs, which utilize solar energy. Chemotrophs can be either autotrophic or heterotrophic.

- Chemoautotrophs (or chemotrophic autotroph), (*Gk*: Chemo = chemical, auto = self, troph = nourishment) in addition to deriving energy from chemical reactions, synthesize all necessary organic compounds from carbon dioxide. Chemoautotrophs use inorganic energy sources, such as hydrogen sulfide, elemental sulfur, ferrous iron, molecular hydrogen, and ammonia. Most are bacteria or archaea that live in hostile environments such as deep sea vents and are the primary producers in such ecosystems. Evolutionary scientists believe that the first organisms to inhabit Earth were chemoautotrophs that produced oxygen as a by-product and later evolved into both aerobic, animal-like organisms and photosynthetic, plant-like organisms. Chemoautotrophs generally fall into several groups: methanogens, halophiles, sulfur oxidizers and reducers, nitrifiers, anammoxbacteria, and thermoacidophiles.
- Chemoheterotrophs (or chemotrophic heterotrophs) (*Gk*: Chemo = chemical, hetero = (an)other, troph = nourishment) are unable to fix carbon and form their own organic compounds. Chemoheterotrophs can be chemolithoheterotrophs, utilizing

inorganic energy sources such as sulfur or chemoorganoheterotrophs, utilizing organic energy sources such as carbohydrates, lipids, and proteins.

Iron and Manganese Oxidizing Bacteria

In the deep oceans, iron oxidizing bacteria derive their energy needs by oxidizing Iron II to Iron III. The extra electron obtained from this reaction powers the cells, replacing or augmenting traditional phototrophism.

- In general, iron oxidizing bacteria can exist only in areas with high iron concentrations, such as new lava beds or areas of hydrothermal activity (where there is dissolved Fe). Most of the ocean is devoid of iron, due to both the oxidative effect of dissolved oxygen in the water and the tendency of prokaryotes to uptake the iron.
- Lava beds supply bacteria with iron straight from the Earth's mantle, but only newly formed igneous rocks have high enough levels of unoxidized iron. In addition, because oxygen is necessary for the reaction, these bacteria are much more common in the upper ocean, where oxygen is more abundant.
- What is still unknown, though, is how exactly iron bacteria extract the iron out of the rock. It is accepted that some mechanism exists that eats away at the rock, perhaps through specialized enzymes or compounds that bring more FeO to the surface. It has been long debated about how much of the weathering of the rock is due to biotic components and how much can be attributed to abiotic components.
- Hydrothermal vents also release large quantities of dissolved iron into the deep ocean, allowing bacteria to survive. In addition, the high thermal gradient around vent systems means a wide variety of bacteria can coexist, each with its own specialized temperature niche.
- Regardless of the catalytic method used, chemoautotrophic bacteria provide a significant but frequently overlooked food source for deep sea ecosystems - which otherwise receive limited sunlight and organic nutrients.

Manganese oxidizing bacteria also make use of igneous lava rocks in much the same way - by oxidizing Mn^{2+} into Mn^{4+}. Manganese is much rarer than iron in oceanic crust, but is much easier for bacteria to extract from the igneous glass. In addition, each manganese oxidation yields around twice the energy as an iron oxidation due to the gain of twice the number of electrons. Much still remains unknown about manganese oxidizing bacteria because they have not been cultured and documented to any great extent.

Phototroph

Phototrophs are the organisms (usually plants) that carry out photosynthesis to acquire energy. They use the energy from sunlight to convert carbon dioxide and water into organic materials to be utilized in cellular functions such as biosynthesis and respiration.

Most phototrophs are autotrophs, also known as photoautotrophs, and can fix carbon. They can be contrasted with chemoautotrophs that obtain their energy by the oxidation of electron donors in their environments. Photoheterotrophs produce ATP through photophosphorylation but use organic compounds to build structures. Some phototrophs are organotrophs, also known as photo-organotrophs.

In an ecological context, phototrophs provide nutrition for all other forms of life (besides other autotrophs such as chemotrophs). In terrestrial environments plants are the predominant variety, while aquatic environments include a range of phototrophic organisms such as algae (e.g. kelp), other protists (such as euglena), phytoplankton and bacteria (such as cyanobacteria).

One product of this process is starch, which is a storage or reserve form of carbon, which can be used when light conditions are too poor to satisfy the immediate needs of the organism. Photosynthetic bacteria have a substance called bacteriochlorophyll, live in lakes and pools, and use the hydrogen from hydrogen sulfide instead of from water, for the chemical process. (The bacteriochlorophyll pigment absorbs light in the extreme UV and infra-red parts of the spectrum which is outside the range used by normal chlorophyll). Cyanobacteria live in fresh water, seas, soil and lichen, and use a plant-like photosynthesis.

A *photolithotrophic autotroph* is an autotrophic organism that uses light energy, and an inorganic electron donor (e.g., H_2O, H_2, H_2S), and CO_2 as its carbon source. Examples include plants. The depth to which sunlight or artificial light can penetrate into water, so that photosynthesis may occur, is known as the photic zone.

In practice, these terms are almost freely combined. Typical examples are as follows:

- chemolithoautotrophs obtain energy from the oxidation of inorganic compounds and carbon from the fixation of carbon dioxide. Examples: Nitrifying bacteria, Sulfur-oxidizing bacteria, Iron-oxidizing bacteria, Knallgas-bacteria

- photolithoautotrophs obtain energy from light and carbon from the fixation of carbon dioxide, using reducing equivalents from inorganic compounds. Examples: Cyanobacteria (water (H_2O) as reducing equivalent donor), Chlorobiaceae, Chromaticaceae (hydrogen sulfide (H_2S) as reducing equivalent donor), *Chloroflexus* (hydrogen (H_2) as reducing equivalent donor)
- chemolithoheterotrophs obtain energy from the oxidation of inorganic compounds, but cannot fix carbon dioxide (CO_2). Examples: some *Thiobacilus*, some *Beggiatoa*, some *Nitrobacter* spp., *Wolinella* (with H_2 as reducing equivalent donor), some Knallgas-bacteria, some sulfate-reducing bacteria
- chemoorganoheterotrophs obtain energy, carbon, and reducing equivalents for biosynthetic reactions from organic compounds. Examples: most bacteria, e.g. *Escherichia coli*, *Bacillus* spp., *Actinobacteria*
- photoorganoheterotrophs obtain energy from light, carbon and reducing equivalents for biosynthetic reactions from organic compounds. Some species are strictly heterotrophic, many others can also fix carbon dioxide and are mixotrophic. Examples: *Rhodobacter*, *Rhodopseudomonas*, *Rhodospirillum*, *Rhodomicrobium*, *Rhodocyclus*, *Heliobacterium*, *Chloroflexus* (alternatively to photolithoautotrophy with hydrogen)

Heterotrophic Microbial Metabolism

Most microbes are heterotrophic (more precisely chemoorganoheterotrophic), using organic compounds as both carbon and energy sources. Heterotrophic microbes live off of nutrients that they scavenge from living hosts (as commensals or parasites) or find in dead organic matter of all kind (saprophages). Microbial metabolism is the main contribution for the bodily decay of all organisms after death. Many eukaryotic microorganisms are heterotrophic by predation or parasitism, properties also found in some bacteria such as *Bdellovibrio* (an intracellular parasite of other bacteria, causing death of its victims) and Myxobacteria such as *Myxococcus* (predators of other bacteria which are killed and lysed by cooperating swarms of many single cells of Myxobacteria). Most pathogenic bacteria can be viewed as heterotrophic parasites of humans or the other eukaryotic species they affect. Heterotrophic microbes are extremely abundant in nature and are responsible for the breakdown of large organic polymers such as cellulose, chitin or lignin which are generally indigestible to larger animals. Generally, the breakdown of large polymers to carbon dioxide (mineralization) requires several different organisms, with one breaking down the polymer into its constituent monomers, one able to use the monomers and excreting simpler waste compounds as by-products, and one able to use the excreted wastes. There are many variations on this theme, as different organisms are able to degrade different polymers and secrete different waste products. Some organisms are even able to degrade more recalcitrant compounds such as petroleum compounds or pesticides, making them useful in bioremediation.

Biochemically, prokaryotic heterotrophic metabolism is much more versatile than that of eukaryotic organisms, although many prokaryotes share the most basic metabolic models with eukaryotes, e.g. using glycolysis (also called EMP pathway) for sugar metabolism and the citric acid cycle to degrade acetate, producing energy in the form of ATP and reducing power in the form of NADH or quinols. These basic pathways are well conserved because they are also involved in biosynthesis of many conserved building blocks needed for cell growth (sometimes in reverse direction). However, many bacteria and archaea utilize alternative metabolic pathways other than glycolysis and the citric acid cycle. A well-studied example is sugar metabolism via the keto-deoxy-phosphogluconate pathway (also called ED pathway) in *Pseudomonas*. Moreover, there is a third alternative sugar-catabolic pathway used by some bacteria, the pentose phosphate pathway. The metabolic diversity and ability of prokaryotes to use a large variety of organic compounds arises from the much deeper evolutionary history and diversity of prokaryotes, as compared to eukaryotes. It is also noteworthy that the mitochondrion, the small membrane-bound intracellular organelle that is the site of eukaryotic energy metabolism, arose from the endosymbiosis of a bacterium related to obligate intracellular *Rickettsia*, and also to plant-associated *Rhizobium* or *Agrobacterium*. Therefore it is not surprising that all mitrochondriate eukaryotes share metabolic properties with these Proteobacteria. Most microbes respire (use an electron transport chain), although oxygen is not the only terminal electron acceptor that may be used. As discussed below, the use of terminal electron acceptors other than oxygen has important biogeochemical consequences.

Fermentation (Biochemistry)

Fermentation is the process of deriving energy from the oxidation of organic compounds, such as carbohydrates, and using an endogenous electron acceptor, which is usually an organic compound. In contrast, respiration is where electrons are donated to an exogenous electron acceptor, such as oxygen, via an electron transport chain. Fermentation is important in anaerobic conditions when there is no oxidative phosphorylation to maintain the production of ATP (Adenosine triphosphate) by glycolysis. During fermentation, pyruvate is metabolised to various different compounds. Homolactic fermentation is the production of lactic acid from pyruvate; alcoholic fermentation is the conversion of pyruvate into ethanol and carbon dioxide; and heterolactic fermentation is the production of lactic acid as well as other acids and alcohols. Fermentation does not necessarily have to be carried out in an anaerobic environment. For example, even in the presence of abundant oxygen, yeast cells greatly prefer fermentation to oxidative phosphorylation, as long as sugars are readily available for consumption.

Sugars are the most common substrate of fermentation, and typical examples of fermentation products are ethanol, lactic acid, and hydrogen. However, more exotic compounds can be produced by fermentation, such as butyric acid and acetone. Yeast carries out fermentation in the production of ethanol in beers, wines and other alcoholic drinks, along with the production of large quantities of carbon dioxide. Fermentation occurs in mammalian muscle during periods of intense exercise where oxygen supply becomes limited, resulting in the creation of lactic acid.

Chemistry

Fermentation products contain chemical energy (they are not fully oxidized), but are considered waste products since they cannot be metabolized further without the use of oxygen (or other more highly-oxidized electron acceptors). A consequence is that the production of adenosine triphosphate (ATP) by fermentation is less efficient than oxidative phosphorylation, whereby pyruvate is fully oxidized to carbon dioxide.

Ethanol Fermentation

Ethanol fermentation (performed by yeast and some types of bacteria) breaks the pyruvate down into ethanol and carbon dioxide. It is important in bread-making, brewing, and wine-making. Usually only one of the products is desired; in bread-making, the alcohol is baked out, and, in alcohol production, the carbon dioxide is released into the atmosphere or used for carbonating the beverage. When the ferment has a high concentration of pectin, minute quantities of methanol can be produced.

The chemical equation below summarizes the fermentation of glucose, whose chemical formula is $C_6H_{12}O_6$. One glucose molecule is converted into two ethanol molecules and two carbon dioxide molecules:

$$C_6H_{12}O_6 \rightarrow 2\ C_2H_5OH + 2\ CO_2$$

C_2H_5OH is the chemical formula for ethanol.

Before fermentation takes place, one glucose molecule is broken down into two pyruvate molecules. This is known as glycolysis.

Lactic Acid Fermentation

Lactic acid fermentation is the simplest type of fermentation. Essentially, it is a redox reaction. In anaerobic conditions, the cell's primary mechanism of ATP production is glycolysis. Glycolysis reduces – transfers electrons to – NAD^+, forming NADH. However, there is only a limited supply of NAD^+ available in a cell. For glycolysis to continue, NADH must be oxidized – have electrons taken away – to regenerate the NAD^+. This is usually done through an electron transport chain in a process called oxidative phosphorylation; however, this mechanism is not available without oxygen.

Instead, the NADH donates its extra electrons to the pyruvate molecules formed during glycolysis. Since the NADH has lost electrons, NAD^+ regenerates and is again available for glycolysis. Lactic acid, for which this process is named, is formed by the reduction of pyruvate.

In heterolactic acid fermentation, one molecule of pyruvate is converted to lactate; the other is converted to ethanol and carbon dioxide. In homolactic acid fermentation, both molecules of pyruvate are converted to lactate. Homolactic acid fermentation is unique because it is one of the only respiration processes to not produce a gas as a byproduct.

Homolactic fermentation breaks down the pyruvate into lactate. It occurs in the muscles of animals when they need energy faster than the blood can supply oxygen. It also occurs in some kinds of bacteria (such as lactobacilli) and some fungi. It is this type of bacteria that converts lactose into lactic acid in yogurt, giving it its sour taste. These lactic acid bacteria can be classed as homofermentative, where the end product is mostly lactate, or heterofermentative, where some lactate is further metabolized and results in carbon dioxide, acetate or other metabolic products.

The process of lactic acid fermentation using glucose is summarized below. In homolactic fermentation, one molecule of glucose is converted to two molecules of lactic acid:

$$C_6H_{12}O_6 \rightarrow 2\ CH_3CHOHCOOH.$$

In heterolactic fermentation, the reaction proceeds as follows, with one molecule of glucose being converted to one molecule of lactic acid, one molecule of ethanol, and one molecule of carbon dioxide:

$$C_6H_{12}O_6 \rightarrow CH_3CHOHCOOH + C_2H_5OH + CO_2$$

Before lactic acid fermentation can occur, the molecule of glucose must be split into two molecules of pyruvate. This process is called glycolysis.

Glycolysis

To extract chemical energy from glucose, the glucose molecule must be split into two molecules of pyruvate. This process also generates two molecules of adenosine triphosphate as an immediate energy yield and two molecules of NADH.

$$C_6H_{12}O_6 + 2\ ADP + 2\ P_i + 2\ NAD^+ \rightarrow 2\ CH_3COCOO^- + 2\ ATP + 2\ NADH + 2\ H_2O + 2H^+$$

The chemical formula of pyruvate is CH_3COCOO^-. P_i stands for the inorganic phosphate. As shown by the reaction equation, glycolysis causes the reduction of two molecules of NAD^+ to NADH. Two ADP molecules are also converted to two ATP and two water molecules via substrate-level phosphorylation.

Aerobic Respiration

In aerobic respiration, the pyruvate produced by glycolysis is further oxidized completely, generating additional ATP and NADH in the citric acid cycle and by oxidative phosphorylation. However, this can only occur in the presence of oxygen. Oxygen is toxic to organisms which are obligate anaerobes, and are not required by facultative anaerobic organisms. In the absence of oxygen, one of the fermentation pathways occurs in order to regenerate NAD^+; lactic acid fermentation is one of these pathways.

Hydrogen Gas Production in Fermentation

Hydrogen gas is produced in many types of fermentation (mixed acid fermentation, butyric acid fermentation, caproate fermentation, butanol fermentation, glyoxylate fermentation), as a way to regenerate NAD^+ from NADH. Electrons are transferred to ferredoxin, which in turn is oxidized by hydrogenase, producing H_2. Hydrogen gas is a substrate for methanogens and sulfate reducers, which keep the concentration of hydrogen sufficiently low to allow the production of such an energy-rich compound.

History

The first solid evidence of the living nature of yeast appeared between 1837 and 1838 when three publications appeared by C. Cagniard de la Tour, T. Swann, and F. Kuetzing, each of whom independently concluded as a result of microscopic investigations that yeast was a living organism that reproduced by budding. The word "yeast," it should be noted, traces its origins back to the Sanskrit word meaning "boiling." It was perhaps because wine, beer, and bread were each basic foods in Europe, that most of the early studies on fermentation were done on yeasts, with which they were made. Soon bacteria were also discovered; the term was first used in English in the late 1840s, but it did not come into general use until the 1870s, and then largely in connection with the new germ theory of disease. Louis Pasteur (1822–1895), during the 1850s and 1860s, showed that fermentation was initiated by living organisms in a series of investigations. In 1857 Pasteur showed that lactic acid fermentation is caused by living organisms. In 1860 he demonstrated that bacteria cause souring in milk, a process formerly thought to be merely a chemical change, and his work in identifying the role of microorganisms in food spoilage led to the process of pasteurization. In 1877, working to improve the French brewing industry, Pasteur published his famous paper on fermentation, Etudes sur la Biere , which was translated into English in 1879 as Studies on Fermentation. He defined fermentation (incorrectly) as "Life without air," but correctly showed specific types of microorganisms cause specific types of fermentations and specific end products.

Although showing that fermentation was generally the result of the action of living microorganisms was a breakthrough, it did not explain the basic nature of the fermentation process, or prove that it was caused by the microorganisms that were apparently always present. Many scientists, including Pasteur, had attempted unsuccessfully to extract the fermentation enzyme from yeast. Success came in 1897 when the German chemist Eduard Buechner ground up yeast, extracted a juice from them, then found to his amazement that this "dead" liquid would ferment a sugar solution, forming carbon dioxide and alcohol much like living yeasts. The "unorganized ferments" behaved just like the organized ones. From that time on the term enzyme came to be applied to all ferments. It was then understood that fermentation is caused by enzymes which are produced by microorganisms. In 1907 Buechner won the Nobel Prize in chemistry for his work.

Advances in microbiology and fermentation technology have continued steadily up until the present. For example, in the late 1970s it was discovered that microorganisms could be mutated with physical and chemical treatments to be higher yielding, faster growing, tolerant of less oxygen, and able to use a more concentrated medium. Strain selection and hybridization developed as well, afecting most modern food fermentations.

Etymology

The word *fermentation* is derived from the Latin verb "fervere" which means "to boil". It is thought to have been first used in the late fourteenth century in alchemy, but only in a broad sense. It was not used in the modern scientific sense until around 1600.

Special Metabolic Properties

Methylotrophy

Methylotrophy refers to the ability of an organism to use C1-compounds as energy sources. These compounds include methanol, methyl amines, formaldehyde, and formate. Several other less common substrates may also be used for metabolism, all of which lack carbon-carbon bonds. Examples of methylotrophs include the bacteria *Methylomonas* and *Methylobacter*. Methanotrophs are a specific type of methylotroph that are also able to use methane (CH_4) as a carbon source by oxidizing it sequentially to methanol (CH_3OH), formaldehyde (CH_2O), formate (HCOO), and carbon dioxide CO_2 initially using the enzyme methane monooxygenase. As oxygen is required for this process, all (conventional) methanotrophs are obligate aerobes. Reducing power in the form of quinones and NADH is produced during these oxidations to produce a proton motive force and therefore ATP generation. Methylotrophs and methanotrophs are not considered as autotrophic, because they are able to incorporate some of the oxidized methane (or other metabolites) into cellular carbon before it is completely oxidized to CO_2 (at the level of formaldehyde), using either the serine pathway (*Methylosinus*, *Methylocystis*) or the ribulose monophosphate pathway (*Methylococcus*), depending on the species of methylotroph.

In addition to aerobic methylotrophy, methane can also be oxidized anaerobically. This occurs by a consortium of sulfate-reducing bacteria and relatives of methanogenic Archaea working syntrophically. Little is currently known about the biochemistry and ecology of this process.

Methanogenesis is the biological production of methane. It is carried out by methanogens, strictly anaerobic Archaea such as

Methanococcus, Methanocaldococcus, Methanobacterium, Methanothermus, Methanosarcina, Methanosaeta and *Methanopyrus*. The biochemistry of methanogenesis is unique in nature in its use of a number of unusual cofactors to sequentially reduce methanogenic substrates to methane, such as coenzyme M and methanofuran.. These cofactors are responsible (among other things) for the establishment of a proton gradient across the outer membrane thereby driving ATP synthesis. Several types of methanogenesis occur, differing in the starting compounds oxidized. Some methanogens reduce carbon dioxide (CO_2) to methane (CH_4) using electrons (most often) from hydrogen gas (H_2) chemolithoautotrophically.

These methanogens can often be found in environments containing fermentative organisms. The tight association of methanogens and fermentative bacteria can be considered to be syntrophic because the methanogens, which rely on the fermentors for hydrogen, relieve feedback inhibition of the fermentors by the build-up of excess hydrogen that would otherwise inhibit their growth. This type of syntrophic relationship is specifically known as interspecies hydrogen transfer. A second group of methanogens use methanol (CH_3OH) as a substrate for methanogenesis. These are chemoorganotrophic, but still autotrophic in using CO_2 as only carbon source. The biochemistry of this process is quite different from that of the carbon dioxide-reducing methanogens. Lastly, a third group of methanogens produce both methane and carbon dioxide from acetate (CH_3COO^-) with the acetate being split between the two carbons. These acetate-cleaving organisms are the only chemoorganoheterotrophic methanogens. All autotrophic methanogens use a variation of the acetyl-CoA pathway to fix CO_2 and obtain cellular carbon.

Syntrophy

Syntrophy, in the context of microbial metabolism, refers to the pairing of multiple species to achieve a chemical reaction that, on its own, would be energetically unfavourable. The best studied example of this process is the oxidation of fermentative end products (such as acetate, ethanol and butyrate) by organisms such as *Syntrophomonas*. Alone, the oxidation of butyrate to acetate and hydrogen gas is energetically unfavourable. However, when a hydrogenotrophic (hydrogen-using) methanogen is present the use of the hydrogen gas will significantly lower the concentration of hydrogen (down to 10^{-5} atm) and thereby shift the equilibrium of the butyrate oxidation reaction under standard conditions ($\Delta G^{o\prime}$) to non-standard conditions ($\Delta G'$). Because the concentration of one product is lowered, the reaction is "pulled" towards the products and shifted towards net energetically favourable conditions (for butyrate oxidation: $\Delta G^{o\prime}$= +48.2 kJ/mol, but $\Delta G'$ = -8.9 kJ/mol at 10^{-5} atm hydrogen and even lower if also the initially produced acetate is further metabolized by methanogens). Conversely, the available free energy from methanogenesis is lowered from $\Delta G^{o\prime}$= -131 kJ/mol under standard conditions to $\Delta G'$ = -17 kJ/mol at 10^{-5} atm hydrogen. This is an example of intraspecies hydrogen transfer. In this way, low energy-yielding carbon sources can be used by a consortium of organisms to achieve further degradation and eventual mineralization of these compounds. These reactions help prevent the excess sequestration of carbon over geologic time scales, releasing it back to the biosphere in usable forms such as methane and CO_2.

Anaerobic Respiration

While aerobic organisms during respiration use oxygen as a terminal electron acceptor, anaerobic organisms use other electron acceptors. These inorganic compounds have a lower reduction potential than oxygen, meaning that respiration is less efficient in these organisms and leads to slower growth rates than aerobes. Many facultative anaerobes can use either oxygen or alternative terminal electron acceptors for respiration depending on the environmental conditions.

Most respiring anaerobes are heterotrophs, although some do live autotrophically. All of the processes described below are dissimilative, meaning that they are used during energy production and not to provide nutrients for the cell (assimilative). Assimilative pathways for many forms of anaerobic respiration are also known.

Denitrification

Denitrification is a microbially facilitated process of nitrate reduction that may ultimately produce molecular nitrogen (N_2) through a series of intermediate gaseous nitrogen oxide products. This respiratory process reduces oxidized forms of nitrogen in response to the oxidation of an electron donor such as organic matter. The preferred nitrogen electron acceptors in order of most to least thermodynamically favourable include nitrate (NO_3^-), nitrite (NO_2^-), nitric oxide (NO), and nitrous oxide (N_2O). In terms of the general nitrogen cycle, denitrification completes the cycle by returning N_2 to the atmosphere. The process is performed primarily by heterotrophic bacteria (such as *Paracoccus denitrificans* and various pseudomonads), although autotrophic denitrifiers have also been identified (e.g., *Thiobacillus denitrificans*). Denitrifiers are represented in all main phylogenetic groups. Generally several species of bacteria are involved in the complete reduction of nitrate to molecular nitrogen, and more than one enzymatic pathway have been identified in the reduction process. Direct reduction from nitrate to ammonium, a process known as dissimilatory nitrate reduction to ammonium or DNRA, is also possible for organisms that have the nrf-gene. This is less common than denitrification in most ecosystems as a means of nitrate reduction. Other genes known in microorganisms which denitrify include *nir* (nitrate reductase) and *nos* (nitrous oxide reductase) among others; organisms identified as having these genes include *Alcaligenes faecalis*, *Alcaligenes xylosoxidans*, many in the *Pseudomonas* genus, *Bradyrhizobium japonicum*, and *Blastobacter denitrificans*.

Nutrient Limitation

All organisms require certain nutrients in their surroundings (available to them) for survival. Depending upon the ecosystem, nitrogen is most likely the limiting nutrient, although phosphorus is the other primary limiting nutrient and these two elements interact chemically. Some organisms appear to be able to denitrify and remove phosphorus. The triple bond of N_2 makes this a very stable compound; most organisms (i.e. plants) depend upon others to break this down to make it available for biochemical reactions.

Conditions Required

Denitrification takes place under special conditions in both terrestrial and marine ecosystems. In general, it occurs where oxygen, a more energetically favourable electron acceptor, is depleted, and bacteria respire nitrate as a substitute terminal electron acceptor. Due to the high concentration of oxygen in our atmosphere, denitrification only takes place in environments where oxygen consumption exceeds the rate of oxygen supply, such as in some soils and groundwater, wetlands, poorly ventilated corners of the ocean, and in seafloor sediments.

Denitrification generally proceeds through some combination of the following intermediate forms:

$$NO_3^- \rightarrow NO_2^- \rightarrow NO + N_2O \rightarrow N_2\ (g)$$

The complete denitrification process can be expressed as a redox reaction:

$$2\ NO_3^- + 10\ e^- + 12\ H^+ \rightarrow N_2 + 6\ H_2O$$

This reaction shows a fractionation in isotope composition. Lighter isotopes of nitrogen are preferred in the reaction, leaving the heavier nitrogen isotopes in the residual matter. The process can cause delta-values of up to “40, where delta is a representation of the difference in isotopic composition. This can be used to identify denitrification processes in nature.

Deliberate use of Process

Denitrification is commonly used to remove nitrogen from sewage and municipal wastewater. It is also an instrumental process in wetlands and riparian zones for the removal of excess nitrate from groundwater resulting from excessive agricultural or residential fertilizer usage.

Reduction under anoxic conditions can also occur through process called anaerobic ammonia oxidation (anammox):

$$NH_4^+ + NO_2^- \rightarrow N_2 + 2\ H_2O$$

In some wastewater treatment plants, small amounts of methanol, ethanol, acetate or proprietary products like MicroCg or MicroCglycerin are added to the wastewater to provide a carbon source for the denitrification bacteria. Denitrification processes are also used in the treatment of industrial wastes.

Influence on Global Climate Change

Increasing carbon dioxide levels within the atmosphere will influence global nutrient cycling, yet it is difficult to predict what those interactions might be. Chemical interactions between soils and the atmosphere will be influenced by changes in atmospheric composition. There are indications that increased fertilization of soils with nitrogen causes a decrease in carbon sequestration.

Jake Beaulieu, a postdoctoral researcher the Environmental Protection Agency in Cincinnati, Ohio and Jennifer Tank, Galla Professor of Biological Sciences at the University of Notre Dame, are lead authors of new paper demonstrating that streams and rivers receiving nitrogen inputs from urban and agricultural land uses are a significant source of nitrous oxide to the atmosphere.

Sulfate Reduction - sulfate as Electron Acceptor

Sulfate reduction is a relatively energetically poor process used by many Gram negative bacteria found within the δ-Proteobacteria, Gram-positive organisms relating to *Desulfotomaculum* or the archaeon *Archaeoglobus*. Hydrogen sulfide (H_2S) is produced as a metabolic end product. For sulfate reduction electron donors and energy are needed.

Electron Donors

Many sulfate reducers are heterotrophic, using carbon compounds such as lactate and pyruvate (among many others) as electron donors, while others are autotrophic, using hydrogen gas (H_2) as an electron donor. Some unusual autotrophic sulfate-reducing bacteria (e.g. *Desulfotignum phosphitoxidans*) can use phosphite (HPO_3^-) as an electron donor whereas others (e.g. *Desulfovibrio sulfodismutans, Desulfocapsa thiozymogenes, Desulfocapsa sulfoexigens*) are capable of sulfur disproportionation (splitting one compound into two different compounds, in this case an electron donor and an electron acceptor) using elemental sulfur (S^0), sulfite ($SO2^{-3}$), and thiosulfate ($S_2O_3^{2-}$) to produce both hydrogen sulfide (H_2S) and sulfate ($SO2^{-4}$).

Energy for Reduction

All sulfate-reducing organisms are strict anaerobes. Because sulfate is energetically stable, before it can be metabolized it must first be activated by adenylation to form APS (adenosine 5'-phosphosulfate) thereby consuming ATP. The APS is then reduced by the enzyme APS reductase to form sulfite ($SO2^{-3}$) and AMP. In organisms that use carbon compounds as electron donors, the ATP consumed is accounted for by fermentation of the carbon substrate. The hydrogen produced during fermentation is actually what drives respiration during sulfate reduction.

Acetogenesis - carbon Dioxide as Electron Acceptor

Acetogenesis

Acetogenesis is a process through which acetate is produced by anaerobic bacteria from a variety of energy (for example, hydrogen) and carbon (for example, carbon dioxide) sources. The different bacterial species that are capable of acetogenesis are collectively termed acetogens.

Chemical Equation

$$2\ CO_2\ (aq) + 4H_2\ (aq) = CH_3COOH\ (aq) + 2H_2O$$

Acetogenesis and Anaerobic Digestion

Acetogenesis is the third stage of anaerobic digestion;

- Hydrolysis: Particulates are solubilised and large polymers are converted into simpler monomers;
- Acidogenesis: Simple monomers are converted into volatile fatty acids;
- Acetogenesis: Volatile fatty acids are converted into acetic acid, carbon dioxide, and hydrogen; and
- Methanogenesis: Acetate is converted into methane and carbon dioxide, while hydrogen is consumed.

Other Inorganic Electron Acceptors

Ferric iron (Fe^{3+}) is a widespread anaerobic terminal electron acceptor both for autotrophic and heterotrophic organisms. Electron flow in these organisms is similar to those in electron transport, ending in oxygen or nitrate, except that in ferric iron-reducing organisms the final enzyme in this system is a ferric iron reductase. Model organisms include *Shewanella putrefaciens* and *Geobacter metallireducens*. Since some ferric iron-reducing bacteria (e.g. *G. metallireducens*) can use toxic hydrocarbons such as toluene as a carbon source, there is significant interest in using these organisms as bioremediation agents in ferric iron-rich contaminated aquifers. Although ferric iron is the most prevalent inorganic electron acceptor, a number of organisms (including the iron-reducing bacteria mentioned above) can use other inorganic ions in anaerobic respiration.

While these processes may often be less significant ecologically, they are of considerable interest for bioremediation, especially when heavy metals or radionuclides are used as electron acceptors. Examples include:

- Manganic ion (Mn^{4+}) reduction to manganous ion (Mn^{2+})
- Selenate (SeO_4^{2-}) reduction to selenite (SeO_3^{2-}) and selenite reduction to inorganic selenium (Se^0)
- Arsenate (AsO_4^{3-}) reduction to arsenite (AsO_3^{3-})
- Uranyl ion ion (UO_2^{2+}) reduction to uranium dioxide (UO_2).

Organic Terminal Electron Acceptors

A number of organisms, instead of using inorganic compounds as terminal electron acceptors, are able to use organic compounds to accept electrons from respiration. Examples include:

- Fumarate reduction to succinate
- Trimethylamine *N*-oxide (TMAO) reduction to trimethylamine (TMA)
- Dimethyl sulfoxide (DMSO) reduction to Dimethyl sulfide (DMS)
- Reductive dechlorination.

TMAO is a chemical commonly produced by fish, and when reduced to TMA produces a strong odor. DMSO is a common marine and freshwater chemical which is also odiferous when reduced to DMS. Reductive dechlorination is the process by which chlorinated organic compounds are reduced to form their non-chlorinated endproducts. As chlorinated organic compounds are often important (and difficult to degrade) environmental polutants, reductive dechlorination is an important process in bioremediation.

Chemolithotrophy

Chemolithotrophy is a type of metabolism where energy is obtained from the oxidation of inorganic compounds. Most chemolithotrophic organisms are also autotrophic. There are two major objectives to chemolithotrophy: the generation of energy (ATP) and the generation of reducing power (NADH).

Hydrogen Oxidation

Many organisms are capable of using hydrogen (H_2) as a source of energy. While several mechanisms of anaerobic hydrogen oxidation have been mentioned previously (e.g. sulfate reducing- and acetogenic bacteria), hydrogen can also be used as an energy source aerobically. In these organisms, hydrogen is oxidized by a membrane-bound hydrogenase causing proton pumping via electron transfer to various quinones and cytochromes. In many organisms, a second cytoplasmic hydrogenase is used to generate reducing power in the form of NADH, which is subsequently used to fix carbon dioxide via the Calvin cycle. Hydrogen-oxidizing organisms, such as *Cupriavidus necator* (formerly *Ralstonia eutropha*), often inhabit oxic-anoxic interfaces in nature to take advantage of the hydrogen produced by anaerobic fermentative organisms while still maintaining a supply of oxygen.

Sulfur Oxidation

Sulfur oxidation involves the oxidation of reduced sulfur compounds (such as sulfide (H_2S), inorganic sulfur (S^0), and thiosulfate ($S_2O_3^{2-}$) to form sulfuric acid (H_2SO_4). A classic example of a sulfur-oxidizing bacterium is *Beggiatoa*, a microbe originally described by Sergei Winogradsky, one of the founders of environmental microbiology. Another example is *Paracoccus*.

Generally, the oxidation of sulfide occurs in stages, with inorganic sulfur being stored either inside or outside of the cell until needed. This two step process occurs because energetically sulfide is a better electron donor than inorganic sulfur or thiosulfate, allowing for a greater number of protons to be translocated across the membrane. Sulfur-oxidizing organisms generate reducing power for carbon dioxide fixation via the Calvin cycle using reverse electron flow, an energy-requiring process that pushes the electrons against their thermodynamic gradient to produce NADH.

Biochemically, reduced sulfur compounds are converted to sulfite (SO_2^{-3}) and subsequently converted to sulfate (SO_2^{-4}) by the enzyme sulfite oxidase. Some organisms, however, accomplish the same oxidation using a reversal of the APS reductase system used by sulfate-reducing bacteria. In all cases the energy liberated is transferred to the electron transport chain for ATP and NADH production. In addition to aerobic sulfur oxidation, some organisms (e.g. *Thiobacillus denitrificans*) use nitrate (NO_3^{2-}) as a terminal electron acceptor and therefore grow anaerobically.

Ferrous Iron (Fe^{2+}) Oxidation

Ferrous iron is a soluble form of iron that is stable at extremely low pHs or under anaerobic conditions. Under aerobic, moderate pH conditions ferrous iron is oxidized spontaneously to the ferric (Fe^{3+}) form and is hydrolyzed abiotically to insoluble ferric hydroxide ($Fe(OH)_3$). There are three distinct types of ferrous iron-oxidizing microbes. The first are acidophiles, such as the bacteria *Acidithiobacillus ferooxidans* and *Leptospirrillum ferrooxidans*, as well as the archaeon *Ferroplasma*. These microbes oxidize iron in environments that have a very low pH and are important in acid mine drainage. The second type of microbes oxidize ferrous iron at cirum-neutral pH. These micro-organisms (for example *Gallionella ferruginea* or *Leptothrix ochracea*) live at the oxic-anoxic interfaces and are microaerophiles. The third type of iron-oxidizing microbes are anaerobic photosynthetic bacteria such as Rhodopsuedomonas, which use ferrous iron to produce NADH for autotrophic carbon dioxide fixation. Biochemically, aerobic iron oxidation is a very energetically poor process which therefore requires large amounts of iron to be oxidized by the enzyme rusticyanin to facilitate the formation of proton motive force. Like sulfur oxidation, reverse electron flow must be used to form the NADH used for carbon dioxide fixation via the Calvin cycle.

Nitrification

Nitrification is the process by which ammonia (NH_3) is converted to nitrate (NO_3^-). Nitrification is actually the net result of two distinct processes: oxidation of ammonia to nitrite (NO_2^-) by nitrosifying bacteria (e.g. *Nitrosomonas*) and oxidation of nitrite to nitrate by the nitrite-oxidizing bacteria (e.g. *Nitrobacter*). Both of these processes are extremely energetically poor leading to very slow growth rates for both types of organisms. Biochemically, ammonia oxidation occurs by the stepwise oxidation of ammonia to hydroxylamine (NH_2OH) by the enzyme ammonia monooxygenase in the cytoplasm, followed by the oxidation of hydroxylamine to nitrite by the enzyme hydroxylamine oxidoreductase in the periplasm.

Electron and proton cycling are very complex but as a net result only one proton is translocated across the membrane per molecule of ammonia oxidized. Nitrite reduction is much simpler, with nitrite being oxidized by the enzyme nitrite oxidoreductase coupled to proton translocation by a very short electron transport chain, again leading to very low growth rates for these organisms. Oxygen is required in both ammonia and nitrite oxidation, meaning that both nitrosifying and nitrite-oxidizing bacteria are aerobes. As in sulfur and iron oxidation, NADH for carbon dioxide fixation using the Calvin cycle is generated by reverse electron flow, thereby placing a further metabolic burden on an already energy-poor process.

Anammox

Anammox stands for anaerobic ammonia oxidation and the organisms responsible were relatively recently discovered, in the late 1990s. This form of metabolism occurs in members of the Planctomycetes (e.g. Candidatus *Brocadia anammoxidans*) and involves the coupling of ammonia oxidation to nitrite reduction. As oxygen is not required for this process these organisms are strict anaerobes. Amazingly, hydrazine (N_2H_4 - rocket fuel) is produced as an intermediate during anammox metabolism. To deal with the high toxicity of hydrazine, anammox bacteria contain a hydrazine-containing intracellular organelle called the anammoxasome, surrounded by highly compact (and unusual) ladderane lipid membrane. These lipids are unique in nature, as is the use of hydrazine as a metabolic intermediate. Anammox organisms are autotrophs although the mechanism for carbon dioxide fixation is unclear. Because of this property, these organisms could be used in industry to remove nitrogen in wastewater treatment processes. Anammox has also been shown have widespread occurrence in anaerobic aquatic systems and has been speculated to account for approximately 50% of nitrogen gas production in the ocean.

Phototrophy

Many microbes (phototrophs) are capable of using light as a source of energy to produce ATP and organic compounds such as carbohydrates, lipids, and proteins. Of these, algae are particularly significant because they are oxygenic, using water as an electron donor for electron transfer during photosynthesis. Phototrophic bacteria are found in the phyla Cyanobacteria, Chlorobi, Proteobacteria, Chloroflexi, and Firmicutes. Along with plants these microbes are responsible for all biological generation of oxygen gas on Earth. Because chloroplasts were derived from a lineage of the Cyanobacteria, the general principles of metabolism in these endosymbionts can also be applied to chloroplasts. In addition to oxygenic photosynthesis, many bacteria can also photosynthesize anaerobically, typically using sulfide (H_2S) as an electron donor to produce sulfate. Inorganic sulfur (S^0), thiosulfate ($S_2O_3^{2-}$) and ferrous iron (Fe^{2+}) can also be used by some organisms. Phylogenetically, all oxygenic photosynthetic bacteria are Cyanobacteria, while anoxygenic photosynthetic bacteria belong to the purple bacteria (Proteobacteria), Green sulfur bacteria (e.g. Chlorobium), Green non-sulfur bacteria (e.g. Chloroflexus), or the heliobacteria (Low %G+C Gram positives). In addition to these organisms, some microbes (e.g. the Archaeon *Halobacterium* or the bacterium *Roseobacter*, among others) can utilize light to produce energy using the enzyme bacteriorhodopsin, a light-driven proton pump. However, there are no known Archaea that carry out photosynthesis.

As befits the large diversity of photosynthetic bacteria, there are many different mechanisms by which light is converted into energy for metabolism. All photosynthetic organisms locate their photosynthetic reaction centres within a membrane, which may be invaginations of the cytoplasmic membrane (Proteobacteria), thylakoid membranes (Cyanobacteria), specialized antenna structures called chlorosomes (Green sulfur and non-sulfur bacteria), or the cytoplasmic membrane itself (heliobacteria). Different photosynthetic bacteria also contain different photosynthetic pigments, such as chlorophylls and carotenoids, allowing them to take advantage of different portions of the electromagnetic spectrum and thereby inhabit different niches. Some groups of organisms contain more specialized light-harvesting structures (e.g. phycobilisomes in Cyanobacteria and chlorosomes in Green sulfur and non-sulfur bacteria), allowing for increased efficiency in light utilization.

Biochemically, anoxygenic photosynthesis is very different from oxygenic photosynthesis. Cyanobacteria (and by extension, chloroplasts) use the Z scheme of electron flow in which electrons eventually are used to form NADH. Two different reaction centres (photosystems) are used and proton motive force is generated both by using cyclic electron flow and the quinone pool. In anoxygenic photosynthetic bacteria, electron flow is cyclic, with all electrons used in photosynthesis eventually being transferred back to the single reaction centre. A proton motive force is generated using only the quinone pool. In heliobacteria, Green sulfur, and Green non-sulfur bacteria, NADH is formed using the protein ferredoxin, an energetically favourable reaction. In purple bacteria, NADH is formed by reverse electron flow due to the lower chemical potential of this reaction centre. In all cases, however, a proton motive force is generated and used to drive ATP production via an ATPase. Most photosynthetic microbes are autotrophic, fixing carbon dioxide via the Calvin cycle. Some photosynthetic bacteria (e.g. Chloroflexus) are photoheterotrophs, meaning that they use organic carbon compounds as a carbon source for growth. Some photosynthetic organisms also fix nitrogen.

Nitrogen Fixation

Nitrogen fixation is the natural process, either biological or abiotic, by which nitrogen (N_2) in the atmosphere is converted into ammonia (NH_3). This process is essential for life because fixed nitrogen is required to biosynthesize the basic building blocks of life, e.g., nucleotides for DNA and RNA and amino acids for proteins. Nitrogen fixation also refers to other abiological conversions of nitrogen, such as its conversion to nitrogen dioxide.

Nitrogen fixation is utilized by numerous prokaryotes, including bacteria, actinobacteria, and certain types of anaerobic bacteria. Microorganisms that fix nitrogen are called diazotrophs. Some higher plants, and some animals (termites), have formed associations (symbioses) with diazotrophs. Nitrogen fixation also occurs as a result of non-biological processes. These include lightning, industrially through the Haber-Bosch Process, and combustion. Biological nitrogen fixation was discovered by the Dutch microbiologist Martinus Beijerinck.

Biological Nitrogen Fixation

Biological nitrogen fixation (BNF) occurs when atmospheric nitrogen is converted to ammonia by an enzyme called nitrogenase. The formula for BNF is:

$$N_2 + 8\,H^+ + 6\,e^- \rightarrow 2\,NH_3 + H_2$$

The process is coupled to the hydrolysis of 16 equivalents of ATP and is accompanied by the co-formation of one molecule of H_2. In free-living diazotrophs, the nitrogenase-generated ammonium is assimilated into glutamate through the glutamine synthetase/glutamate synthase pathway.

Enzymes responsible for nitrogenase action are very susceptible to destruction by oxygen. (In fact, many bacteria cease production of the enzyme in the presence of oxygen). Many nitrogen-fixing organisms exist only in anaerobic conditions, respiring to draw down oxygen levels, or binding the oxygen with a protein such as Leghemoglobin.

Plants that contribute to nitrogen fixation include the legume family – Fabaceae – with taxa such as clover, soybeans, alfalfa, lupines, peanuts, and rooibos. They contain symbiotic bacteria called *Rhizobia* within nodules in their root systems, producing nitrogen compounds that help the plant to grow and compete with other plants. When the plant dies, the fixed nitrogen is released, making it available to other plants and this helps to fertilize the soil The great majority of legumes have this association, but a few genera (e.g., *Styphnolobium*) do not. In many traditional and organic farming practices, fields are rotated through various types of crops, which usually includes one consisting mainly or entirely of clover or buckwheat (family *Polygonaceae*), which were often referred to as "green manure."

Non-leguminous Nitrogen-fixing Plants

Although by far the majority of nitrogen-fixing plants are in the legume family Fabaceae, there are a few non-leguminous plants, such as alder, that can also fix nitrogen. These plants, referred to as "actinorhizal plants", consist of 24 genera of woody shrubs or trees distributed among in 8 plant families. The ability to fix nitrogen is not universally present in these families. For instance, of 122 genera in the Rosaceae, only 4 genera are capable of fixing nitrogen. All these families belong to the orders Cucurbitales, Fagales, and Rosales, which together with the Fabales form a clade of eurosids. There are also several nitrogen-fixing symbiotic associations that involve cyanobacteria (such as *Nostoc*):

- Some lichens such as *Lobaria* and *Peltigera*
- Mosquito fern (*Azolla* species)
- Cycads
- *Gunnera*.

Microorganisms that Fix Nitrogen (Diazotrophs)

- Cyanobacteria
- *Azotobacteraceae*
- *Rhizobia*
- *Frankia*.

Nitrogen Fixation by Cyanobacteria

Cyanobacteria inhabit nearly all illuminated environments on Earth and play key roles in the carbon and nitrogen cycle of the biosphere. In general, cyanobacteria are able to utilize a variety of inorganic and organic sources of combined nitrogen, like nitrate, nitrite, ammonium, urea, or some amino acids. Several cyanobacterial strains are also capable of diazotrophic growth. Genome sequencing has provided a large amount of information on the genetic basis of nitrogen metabolism and its control in different cyanobacteria. Comparative genomics, together with functional studies, has led to a significant advance in this field over the past years. 2-Oxoglutarate has turned out to be the central signalling molecule reflecting the carbon/nitrogen balance of cyanobacteria. Central players of nitrogen control are the global transcriptional factor NtcA, which controls the expression of many genes involved in nitrogen metabolism, as well as the P_{II} signalling protein, which fine-tunes cellular activities in response to changing C/N conditions. These two proteins are sensors of the cellular 2-oxoglutarate level and have been conserved in all cyanobacteria. In contrast, the adaptation to nitrogen starvation involves heterogeneous responses in different strains. Nitrogen fixation by cyanobacteria in coral reefs can fix twice the amount of nitrogen than on land–around 1.8 kg of nitrogen is fixed per hectare per day.

Chemical Nitrogen Fixation

Nitrogen can also be artificially fixed for use in fertilizers, explosives, or in other products. The most common method is the Haber process. Artificial fertilizer production is now the largest source of human-produced fixed nitrogen in the Earth's ecosystem.

The Haber process requires high pressures (around 200 atm) and high temperatures (at least 400 °C), routine conditions for industrial catalysis. This highly efficient process uses natural gas as a hydrogen source and air as a nitrogen source.

Research on Catalytic Nitrogen-fixation

Much research has been conducted on the discovery of catalysts for nitrogen fixation, often with the goal of reducing the energy required for this conversion. However, such research has thus far failed to even approach the efficiency and ease of the Haber process. Many compounds react with atmospheric nitrogen under ambient conditions. For example, lithium metal converts to lithium nitride under an atmosphere of nitrogen. Treatment of the resulting nitride gives ammonia.

The first dinitrogen complex was reported in 1965 based on ammonia coordinated to ruthenium ($[Ru(NH_3)_5(N_2)]^{2+}$), research in chemical fixation focused on transition metal complexes. Since then, a large number of transition metal compounds that contain dinitrogen as a ligand have been discovered. The dinitrogen ligand can either be bound to a single metal or bridge two (or more) metals. The coordination chemistry of dinitrogen is complex and currently under intense investigation. This research may lead to new ways of using dinitrogen in synthesis and on an industrial scale.

The first example of homolytic cleavage of dinitrogen under mild conditions was published in 1995. Two equivalents of a molybdenum complex reacted with one equivalent of dinitrogen, creating a triple bonded MoN complex. Since then, this triple bounded complex has been used to make nitriles. The first catalytic system converting nitrogen to ammonia at room temperature and pressure was discovered in 2003 and is based on another molybdenum compound, a proton source, and a strong reducing agent. However, this catalytic reduction fixates only a few nitrogen molecules.

Iron in Microbiology

Iron is essential for almost all living organisms as it is involved in a wide variety of important metabolic processes. However, iron is not readily available and microorganisms therefore employ various iron uptake systems to secure sufficient supplies from their surroundings. There is considerable variation in the range of iron transporters and iron sources utilised by different microbial species. Pathogens, in particular, require efficient iron acquisition mechanisms to enable them to compete successfully for iron in the highly iron-restricted environment of the host's tissues and body fluids.

Siderophores

In the early days of life on earth due to the emergence of photosynthesis, the atmosphere became oxidative and so iron prevailed in its trivalent form. The consequence was that most microorganisms live today in a surrounding where the concentration of soluble iron compounds is much too low to satisfy their physiological needs. One of the possibilities to circumvent this problem is the production of so-called siderophores, compounds which can form water soluble complexes with Fe^{3+}. They are released in situations of iron deficiency. Fe^{3+} possesses six coordination sites arranged in the shape of an octahedron which can accommodate three bidentate ligands. The major ligand types are catecholates, hydroxamates and alpha-hydroxycarboxylates, in the ideal case connected by adequate spacers which keep them in the correct position. Many siderophores have elaborate structures which can bind to the cell surface of the producing species. In this way pirating by competing microorganisms is rendered more difficult.

Heme Uptake and Iron Extraction by Bacteria

Heme is ubiquitous, abundant and necessary for energy metabolism. Most bacteria have a heme biosynthesis pathway, but nevertheless, since heme is a major source of iron (an essential metal), microbes take up exogenous heme to retrieve iron. To grab heme, microbes extract it from host hemoproteins. This is achieved by two non-exclusive distinct pathways. One pathway involves proteins secreted by bacteria (hemophores) that scavenge heme from host hemoproteins. The second pathway involves microbial cell surface receptors that catch hemoproteins circulating in the vicinity of the cell surface. Both pathways lead to heme docking to cell surface receptors. In Gram-negative bacteria, docked heme is transported through the outer membrane by an energy-dependent process. In Gram-positive bacteria, docked heme is transferred to membrane-anchored heme binding lipoproteins. In all thus far described systems, heme is actively transported through the plasma membrane by an ATP hydrolysis-powered ABC transporter. Heme is either degraded into biliverdin, CO and iron by heme oxygenases, or iron is retrieved from heme, keeping the tetrapyrrol ring intact by recently identified enzymes. As excess heme is toxic, heme uptake, efflux and degradation are usually highly regulated. In most cases, intracytoplasmic heme or iron released during heme degradation are cofactors along with transcriptional regulators. In several cases, heme uptake and efflux are regulated by extracellular heme.

Rhizobia

The rhizobia live as free-living soil bacteria or in symbiosis with leguminous plants. The success of these organisms in each milieu involves the ability to sense the environment to assess the availability of nutrients, and to optimize cellular systems for their acquisition. Iron in the rhizosphere is mostly inaccessible due to low solubility, and microorganisms must compete for this limited nutrient. Rhizobia belong to the alpha-Proteobacteria, a diverse taxonomic group that includes numerous species that form close or intracellular associations with eukaryotic hosts in a symbiotic or pathogenic context. Thus, in addition to their agricultural and economic importance, rhizobia are model organisms that have given new insights into related, but less tractable animal pathogens. In particular, genetic control of iron homeostasis in the rhizobia and other alpha-Proteobacteria has moved away from the Fur paradigm to an iron sensing mechanism responding to the metal indirectly. Moreover, utilization of heme as an iron source is not unique to animal pathogens, but is an acquisition strategy employed by the rhizobia with some interesting novel features.

Bordetella

Upon colonization of the mammalian respiratory epithelium by mucosal pathogens of the genus *Bordetella*, the host-pathogen interaction causes inflammatory changes, immune activation, and host cell injury. In this dynamic environment, *Bordetella* cells scavenge the nutritional iron necessary for growth. The three classical *Bordetella* species produce the siderophore alcaligin.

In addition, they can utilize xenosiderophores that could be produced by commensals or other microbes that transiently inhabit the nasopharynx. As infection progresses, extravasation of immune cells, erythrocytes and serum to the mucosal surface can occur, exacerbated by the damaging action of *Bordetella* toxins, thus providing iron sources such as transferrin and heme compounds to the microbe. The three characterized *Bordetella* iron systems for utilization of alcaligin, enterobactin and heme are each inducible by the cognate iron source. The ability to sense and respond to the presence of available iron sources allows these pathogens to adapt to temporal changes in iron source availability, and this ability is important for successful *in vivo* growth.

Iron Shigella and E. Coli

Shigella spp. and pathogenic *E. coli* are characterized by a variety and abundance of iron transport systems. Although members of this group of bacteria are closely related genetically, they differ widely in the iron transport systems they use. This may reflect the different niches occupied by different strains and the nature of the source of iron available in a specific environment.

Only the ferrous iron transporter Feo is common to all the commensals and pathogens. All members of this group produce one or more siderophore, but no single siderophore is produced by all. Other iron transport systems include heme transporters and the ferrous iron transporters Sit and Efe. With the exception of the genes for enterobactin and the Feo system, the iron transport genes in the enterics are found within pathogenicity islands or on plasmids and their presence often increases pathogenicity or colonization of niches within the host.

Iron in Erwinia

The critical role of iron in host-pathogen relationships has been elucidated in infectious diseases of mammals, where the importance of siderophores in microbial pathogenesis has been demonstrated. Our group has established the role of iron and its ligands in the virulence of the plant pathogenic bacteria *Dickeya dadantii* (*Erwinia chrysanthemi*) and *Erwinia amylovora*. The genomes of the two pectinolytic enterobacterial species *Pectobacterium atrosepticum* SCRI1043 and *D. dadantii* 3937 have been sequenced and annotated. This review focuses on the functions involved in iron acquisition in both species. Besides the production and utilization of siderophores, *P. atrosepticum* and *D. datantii* have the capacity to use other iron sources. Indeed, both species are able to use haem iron, whereas only *P. atrosepticum* can transport the ferric citrate complex and only *D. dadantii* can acquire ferrous iron. These different modes of iron capture indicate that these species have to cope with various environmental and ecological conditions during their pathogenic life cycle.

Vibrio and *Aeromonas*

Vibrio and *Aeromonas* species are ubiquitous bacteria in aquatic environments worldwide. Many of the species are important pathogens for humans and/or aquatic animals. Several iron acquisition strategies have been developed by vibrios and aeromonads in order to get this essential element for surviving in their host and in aquatic habitats. All species studied so far have the ability to synthesize siderophores to sequester iron from the cell environment and transport it through their respective cognate outer membrane receptors.

It has been demonstrated that this capacity is a relevant virulence factor for human and animal pathogens. Furthermore, all species studied can utilize exogenous siderophores, made by other bacteria. Another iron acquisition system described in both genera involves the use of heme as a source of iron, by a mechanism very well conserved among all species, which involves a heme transporter that includes a specific TonB-dependent outer membrane receptor(s) and an ABC-type inner membrane transporter. Alternative systems based on ferrous or ferric iron transporters have been reported in *V. cholerae*. How the different iron acquisition systems work together to supply iron to the cell and how they are used in the different environments where vibrios and aeromonads can be found is still an open question.

Francisella

Francisella tularensis is unusual among Gram-negative bacteria in that its genome does not encode orthologs for TonB, ExbB and ExbD that typically energize the uptake of iron across the outer membrane. This organism secretes however a siderophore similar in structure to rhizoferrin. The *fsl* operon of six genes encodes functions for biosynthesis and uptake of the siderophore. Two of these genes encode a siderophore synthetase belonging to the nonribosomal peptide synthetase (NRPS)-independent synthetase (NIS)-family and a protein belonging to the pyridoxyl phosphate-dependent decarboxylase family, and both are required for siderophore production. Siderophore utilization involves the product of the *fslE* gene, a protein unique to *Francisella* species that could function as a siderophore receptor. Additionally, genes related in sequence to *fslE* also play a role in siderophore acquisition. The mechanism for TonB-independent iron uptake in this microorganism remains to be elucidated.

Bacteroides

Bacteroides spp. have an essential requirement for heme and non-heme iron. They cannot synthesize the tetrapyrrole macrocycle ring due to a lack of genes for the heme biosynthetic pathway. It is remarkable that heme-dependent organisms outnumber heme-independent organisms in the lower intestinal tract suggesting that heme biosynthesis is not essential for colonization of the colonic environment. However, this colonization advantage may be due to the fact that under anaerobic conditions in the presence of heme, *B. fragilis* can generate nearly the double amount of ATP than *Escherichia coli* per mol of glucose. This high energy yield is linked to a rudimentary heme-induced fumarate reductase and cytochrome *b*-dependent electron transport energy metabolism pathway which uses fumarate as the terminal electron acceptor. Moreover, *Bacteroides* spp. can incorporate iron-deuteroporphyrin and iron-mesoporphyrin into a functional type-*b* cytochrome. Heme can be demetalated without cleaving the tetrapyrrole ring releasing free iron and free protoporphirin IX. The ability of the opportunistic human pathogen *B. fragilis* to cause infections seems to be due in part to its ability to scavenge heme and iron from host proteins. The in-frame translated intergenic region of the fused FeoAB proteins are exclusively present in gastro-intestinal colonizers belonging to the Bacteroidetes, Firmicutes and Actinobacteria phyla. Several members of the *Bacteroides* group have three orthologs of the mammalian-type bacterial ferritin gene, *ftnA*. FtnA may play an important role in protection against iron-induced oxidative stress in this group of highly aerotolerant anaerobes.

Campylobacter

Iron is known to catalyse a wide range of biochemical reactions essential for most living organisms, including *Campylobacter jejuni*. Paradoxically, this iron reactivity is also responsible for the generation of hydroxyl radicals (·OH), which are particularly biotoxic. In order to avoid iron toxicity, microorganisms must achieve an effective iron homeostasis by tightly regulating the expression of genes encoding the proteins involved in iron acquisition, metabolism and oxidative stress defences in response to iron availability. Interestingly, in addition to the classical ferric uptake regulator Fur, *C. jejuni* carries another member of the Fur family of metalloregulators, PerR. PerR is a peroxide-sensing regulator and typically regulates peroxide stress response in Gram-positive bacteria. Recent work indicates that the regulatory functions of Fur and PerR extend beyond their classically ascribed roles. These diverse functions include energy metabolism, protein glycosylation and flagella biogenesis. Moreover, the Fur and PerR regulons appear to overlap and co-regulate key genes at specific junctions.

Cyanobacteria

Cyanobacteria are dependent on but can also be compromised by metals such as iron. On the one hand the demand for iron for photosystem functionality represents a challenge for the iron uptake machinery in iron limiting environments. On the other hand intoxication by iron causes a severe problem for growth and reproduction. To overcome this dilemma cyanobacteria have developed a regulatory network controlling iron uptake. They produce siderophores, which are distinct from that of other bacteria. Furthermore, the iron metabolism is linked to the nitrogen metabolism as documented for example in *Anabaena* sp. PCC 7120.

Bacillus

Bacillus subtilis is a metabolically versatile soil microbe and Gram-positive model organism that displays a sophisticated adaptive response to conditions of iron limitation. The endogenous siderophore of *B. subtilis* is bacillibactin, a trimeric catecholate siderophore similar in structure to enterobactin. In addition to bacillibactin, *B. subtilis* can obtain iron from several xenosiderophores, ferric citrate, heme, and through a newly discovered elemental iron permease.

The regulation of iron homeostasis in *B. subtilis* is complex and involves a ferric uptake regulator (Fur) protein as master regulator and at least two subsidiary regulatory systems.

The most significant of these is an iron-sparing/prioritization response controlled by the small RNA FsrA and three auxiliary proteins (FbpABC). In addition, the bacillibactin uptake system is transcriptionally activated by an AraC family activator, Btr that directly senses bacillibactin. Iron uptake and homeostasis systems in *B. anthracis* and related organisms are largely similar to those in *B. subtilis* with some additional components. These include a second siderophore synthesis operon for petrobactin, which is important for virulence, and a more elaborate (or at least better understood) heme uptake system.

Staphylococci

Staphylococcus aureus causes a significant amount of human morbidity and mortality. The ability of *S. aureus* to cause disease is dependent upon its acquisition of iron from the host. *S. aureus* can obtain iron from various sources during infection, including heme and transferrin. The most abundant iron source in humans is heme-iron bound by hemoglobin contained within erythrocytes. *S. aureus* is known to lyse erythrocytes through secretion of pore-forming toxins, providing access to host hemoglobin. Proteins of the iron-regulated surface determinant (Isd) system bind host hemoproteins, remove the heme cofactor, and shuttle heme into the cytoplasm for use as a nutrient iron source. Deletion of Isd system components decreases staphylococcal virulence, underscoring the importance of heme-iron acquisition during infection. In addition to heme, *S. aureus* can utilize transferrin-iron through the secretion of siderophores. Several staphylococcal siderophores have been described, some of which have defined roles during the pathogenesis of staphylococcal infections. A greater understanding of staphylococcal iron acquisition may lead to the development of novel therapeutic strategies that target nutrient uptake and decrease the threat of this increasingly drug-resistant bacterial pathogen.

Yeasts

Yeasts take up iron by three main mechanisms. In the reductive uptake mechanism, specialized flavo-hemoproteins (Fre) dissociate extracellular ferric complexes by reduction involving trans-plasma membrane electron transfer.

The resulting free iron is then imported by a high-affinity permease system (Ftr), coupled to a copper-dependent oxidase (Fet), which channels iron through the plasma membrane. As a consequence, iron uptake by this mechanism is dependent on the availability of copper. In the siderophore-mediated mechanism, siderophores excreted by the cells or produced by other bacterial or fungal species are taken up without prior dissociation, via specific, copper-independent high-affinity receptors.

The iron is then dissociated from the siderophores intracellularly, probably by reduction. In the heme uptake mechanism, free heme or heme bound to hemoglobin is taken up as such, probably by endocytosis.

Iron is released intracellularly after hydrolysis of the porphyrin ring catalysed by heme oxygenase. Within the cell, iron is stored in vacuoles or in siderophores. Iron can be mobilized from vacuoles by a reductive mechanism homologous to that found at the plasma membrane. Regulation of iron uptake and iron use are mediated by transcriptional regulators acting either as activators in iron-deficient conditions or as repressors in iron-rich conditions, according to the yeast species; these regulators thus adjust the iron uptake flux to the cell's requirements. In the baker's yeast, *Saccharomyces cerevisiae*, a post-transcriptional mechanism is active under low iron conditions, involving the degradation of RNAs encoding inessential iron-utilizing proteins. Other fungi have mechanisms serving a similar purpose at the transcriptional level. Studies in *S. cerevisiae* show that mitochondria are central to regulating cellular iron homeostasis, through the synthesis of iron-sulfur clusters.

Environmental Microbiology

Environmental microbiology is the study of the composition and physiology of microbial communities in the environment. The environment in this case means the soil, water, air and sediments covering the planet and can also include the animals and plants that inhabit these areas. Environmental microbiology also includes the study of microorganisms that exist in artificial environments such as bioreactors.

Microbial life is amazingly diverse and microorganisms literally cover the planet.

An average gram of soil contains approximately one billion (1,000,000,000) microbes representing probably several thousand species. Microorganisms have special impact on the whole biosphere. They are the backbone of ecosystems of the zones where light cannot approach. In such zones, chemosynthetic bacteria are present which provide energy and carbon to the other organisms there. Some microbes are decomposers which have ability to recycle the nutrients. Microbes have a special role in biogeochemical cycles. Microbes, especially bacteria, are of great importance because their symbiotic relationship (either positive or negative) have special effects on the ecosystem.

Microorganisms are used for *in-situ* microbial biodegradation or bioremediation of domestic, agricultural and industrial wastes and subsurface pollution in soils, sediments and marine environments. The ability of each microorganism to degrade toxic waste depends on the nature of each contaminant. Since most sites typically have multiple pollutant types, the most effective approach to microbial biodegradation is to use a mixture of bacterial species and strains, each specific to the biodegradation of one or more types of contaminants. It is vital to monitor the composition of the indigenous and added bacteria in order to evaluate the activity level and to permit modifications of the nutrients and other conditions for optimizing the bioremediation process.

Oil Biodegradation

Petroleum oil is toxic, and pollution of the environment by oil causes major ecological concern. Oil spills of coastal regions and the open sea are poorly containable and mitigation is difficult; much of the oil can, however, be eliminated by the hydrocarbon-degrading activities of microbial communities, in particular the hydrocarbonoclastic bacteria (HCB). These organisms can help remedy the ecological damage caused by oil pollution of marine habitats. HCB also have potential biotechnological applications in the areas of bioplastics and biocatalysis.

Waste Biotreatment

Biotreatment, the processing of wastes using living organisms, is an environmentally friendly alternative to other options for treating waste material. Bioreactors have been designed to overcome the various limiting factors of biotreatment processes in highly controlled systems. This versatility in the design of bioreactors allows the treatment of a wide range of wastes under optimized conditions. It is vital to consider various microorganisms and a great number of analyses are often required.

Wastewater Treatment

Wastewater treatment processes are geared towards one purpose: cleaning up water. Recent application of molecular techniques is unveiling the microbial composition and architecture of the complex communities involved in the treatment processes. It is now recognized that wastewater processes harbor a vast variety of microorganisms most of which are yet-to-be cultured, hence uncharacterized. Metagenomic technology is being used to study the diversity, structure and functions of microbial communities in nitrifying processes, anaerobic ammonia oxidation processes and methane fermenting processes.

Environmental Genomics of Cyanobacteria

The application of molecular biology and genomics to environmental microbiology has led to the discovery of a huge complexity in natural communities of microbes. Diversity surveying, community fingerprinting and functional interrogation of natural populations have become common, enabled by a range of molecular and bioinformatics techniques. Recent studies on the ecology of Cyanobacteria have covered many habitats and have demonstrated that cyanobacterial communities tend to be habitat-specific and that much genetic diversity is concealed among morphologically simple types. Molecular, bioinformatics, physiological and geochemical techniques have combined in the study of natural communities of these bacteria.

Corynebacteria

Corynebacteria are a diverse group Gram-positive bacteria found in a range of different ecological niches such as soil, vegetables, sewage, skin, and cheese smear. Some, such as *Corynebacterium diphtheriae*, are important pathogens while others, such as *Corynebacterium glutamicum*, are of immense industrial importance. *C. glutamicum* is one of the biotechnologically most important bacterial species with an annual production of more than two million tons of amino acids, mainly L-glutamate and L-lysine.

Legionella

Legionella is common in many environments, with at least 50 species and 70 serogroups identified. *Legionella* is commonly found in aquatic habitats where its ability to survive and to multiply within different protozoa equips the bacterium to be transmissible and pathogenic to humans.

Archaea

Originally, Archaea were once thought of as extremophiles existing only in hostile environments but have since been found in all habitats and may contribute up to 20% of total biomass. Archaea are particularly common in the oceans, and the archaea in plankton may be one of the most abundant groups of organisms on the planet. Archaea are subdivided into four phyla of which two, the Crenarchaeota and the Euryarchaeota, are most intensively studied.

Lactobacillus

Lactobacillus species are found in the environment mainly associated with plant material. They are also found in the gastrointestinal tract of humans, where they are symbiotic and make up a portion of the gut flora.

Aspergillus

Aspergillus spores are common components of aerosols where they drift on air currents, dispersing themselves both short and long distances depending on environmental conditions. When the spores come in contact with a solid or liquid surface, they are deposited and if conditions of moisture are right, they germinate. The ability to disperse globally in air currents and to grow almost anywhere when appropriate food and water are available means that ubiquitous is among the most common adjectives used to describe these moulds.

Rhizobia

Symbiotic nitrogen fixation is a mutualistic process in which bacteria reside inside plants and reduce atmospheric nitrogen to ammonia. This ammonia can then be used by the plant for the synthesis of proteins and other nitrogen-containing compounds such as nucleic acids. The Gram-negative soil bacteria that carry out this process are collectively referred to as rhizobia (from the Greek words Riza = Root and Bios = Life). The process of symbiotic nitrogen fixation is of agricultural and ecological significance because plants capable of nitrogen fixation do not need to compete for limited quantities of soil nitrogen, nor do they require expensive nitrogenous fertilizers that can be harmful to the environment.

Microalgae

Algae are a highly diverse group of protists, ranging from simple, unicellular organisms to complex, multicellular entities with a range of differentiated tissues and distinct organs. They are found among diverse aquatic ecosystems and play important roles by supplying carbon and energy as well as providing habitat to other members of the biological communities. Some algae cause significant environmental and health problems. There are three algal groups: the dinoflagellates, the diatoms and the haptophytes. Their 3 main phylla are chlorophyta,rhodophyta and phaeophyta.

Anaerobic Protozoa

Diplomonads are a group of mitochondrion-lacking, binucleated flagellates found in anaerobic or micro-aerophilic environments. Most research on diplomonads has focused on *Giardia*, which is a major cause of water-borne enteric disease in humans and other animals. The first diplomonad to have its genome sequenced was a *Giardia* isolate (WB) and the 11.7 million basepair genome is compact in structure and content with simplified basic cellular machineries and metabolism.

Water Microbiology

An adequate supply of safe drinking water is one of the major prerequisites for a healthy life, but waterborne diseases are still a major cause of death in many parts of the world, particularly in young children, the elderly, or those with compromised immune systems. As the epidemiology of waterborne diseases is changing, there is a growing global public health concern about new and reemerging infectious diseases that are occurring through a complex interaction of social, economic, evolutionary, and ecological factors. An important challenge is therefore the rapid, specific and sensitive detection of waterborne pathogens. Presently, microbial tests are based essentially on time-consuming culture methods. However, newer enzymatic, immunological and genetic methods are being developed to replace and/or support classical approaches to microbial detection. Moreover, innovations in nanotechnology and nanosciences are having a significant impact in biodiagnostics, where a number of nanoparticle-based assays and nanodevices have been introduced for biomolecular detection. Molecular techniques based on genomics, proteomics and transcriptomics are rapidly growing as complete microbial genome sequences are becoming available, and advances are made in sequencing technology, analytical biochemistry, microfluidics and data analysis. While the clinical and food industries are increasingly adapting these techniques, there appear to be major challenges in detecting health-related microbes in source and treated drinking waters. This is due in part to the low density of pathogens in water, necessitating significant processing of large volume samples. From the vast panorama of available molecular techniques, some are finding a place in the water industry: Quantitative PCR, protein detection and immunological approaches, loop-mediated isothermal amplification (LAMP), microarrays.

Sensory Mechanisms

Bacteria have evolved abilities to regulate aspects of their behaviour (such as gene expression) in response to signals in the intracellular and extracellular environment. The interaction of a signal with its receptor (usually a protein or RNA molecule) triggers a series of events that lead to reprogramming of cellular physiology, typically as a consequence of altered patterns of gene expression. In this way, the bacterial cell is able to mount appropriate and effective responses to changing physical and/or chemical environments. The versatility with which many bacteria adapt to environmental change underlies many important aspects of microbiology. For example, pathogens encounter multiple environments as they invade a host from the outside, and then progress through different sites within host tissues. There is growing evidence that pathogenic bacteria make use of physical and chemical cues to signal their presence in a suitable host, and need to adapt to the host environment in order to mount a successful infection. On the other hand, it should not be assumed that all signals to which bacteria must respond originate in the extracellular environment. For many species, even the cosseted life in a laboratory shake flask is 'stressful', in the sense that there is often a need to avoid or reverse the effects of harmful intermediates or by-products of metabolism. For example, all organisms that use dioxygen as a terminal electron acceptor have to deal with the reactive oxygen species that arise as adventitious by-products of aerobic metabolism. In bacteria, multiple protein receptors for oxygen radicals have been described, which control the expression of genes encoding enzymes that detoxify oxygen radicals or repair the damage that they cause.

Bioremediation

Bioremediation is the use of microorganismal metabolism to remove pollutants. Technologies can be generally classified as *in situ* or *ex situ*. *In situ* bioremediation involves treating the contaminated material at the site, while *ex situ* involves the removal of the contaminated material to be treated elsewhere. Some examples of bioremediation technologies are bioventing, landfarming, bioreactor, composting, bioaugmentation, rhizofiltration, and biostimulation.

Bioremediation can occur on its own (natural attenuation or intrinsic bioremediation) or can be spurred on via the addition of fertilizers to increase the bioavailability within the medium (biostimulation). Recent advancements have also proven successful via the addition of matched microbe strains to the medium to enhance the resident microbe population's ability to break down contaminants. Microorganisms used to perform the function of bioremediation are known as Bioremediators.

Not all contaminants, however, are easily treated by bioremediation using microorganisms. For example, heavy metals such as cadmium and lead are not readily absorbed or captured by organisms. The assimilation of metals such as mercury into the food chain may worsen matters. Phytoremediation is useful in these circumstances because natural plants or transgenic plants are able to bioaccumulate these toxins in their above-ground parts, which are then harvested for removal. The heavy metals in the harvested biomass may be further concentrated by incineration or even recycled for industrial use.

The elimination of a wide range of pollutants and wastes from the environment requires increasing our understanding of the relative importance of different pathways and regulatory networks to carbon flux in particular environments and for particular compounds, and they will certainly accelerate the development of bioremediation technologies and biotransformation processes.

Genetic Engineering Approaches

The use of genetic engineering to create organisms specifically designed for bioremediation has great potential. The bacterium *Deinococcus radiodurans* (the most radioresistant organism known) has been modified to consume and digest toluene and ionic mercury from highly radioactive nuclear waste.

Most commonly, the process is misunderstood. The microbes are ever-present in any given context and are generally referred to as "normal microbial flora". During bioremediation (biodegradation) processes, fertilizer/nutrient supplementation is introduced to the environments in efforts to maximize growth and production potential. Common misbelief is that microbes are transported and dispersed into an unadulterated environment.

Mycoremediation

Mycoremediation is a form of bioremediation in which fungi are used to decontaminate the area. The term *mycoremediation* refers specifically to the use of fungal mycelia in bioremediation.

One of the primary roles of fungi in the ecosystem is decomposition, which is performed by the mycelium. The mycelium secretes extracellular enzymes and acids that break down lignin and cellulose, the two main building blocks of plant fibre. These are organic compounds composed of long chains of carbon and hydrogen, structurally similar to many organic pollutants. The key to mycoremediation is determining the right fungal species to target a specific pollutant. Certain strains have been reported to successfully degrade the nerve gases VX and sarin.

In one conducted experiment, a plot of soil contaminated with diesel oil was inoculated with mycelia of oyster mushrooms; traditional bioremediation techniques (bacteria) were used on control plots. After four weeks, more than 95% of many of the PAH (polycyclic aromatic hydrocarbons) had been reduced to non-toxic components in the mycelial-inoculated plots. It appears that the natural microbial community participates with the fungi to break down contaminants, eventually into carbon dioxide and water. Wood-degrading fungi are particularly effective in breaking down aromatic pollutants (toxic components of petroleum), as well as chlorinated compounds (certain persistent pesticides; Battelle, 2000).

Mycofiltration is a similar process, using fungal mycelia to filter toxic waste and microorganisms from water in soil.

Advantages

There are a number of cost/efficiency advantages to bioremediation, which can be employed in areas that are inaccessible without excavation. For example, hydrocarbon spills (specifically, petrol spills) or certain chlorinated solvents may contaminate groundwater, and introducing the appropriate electron acceptor or electron donor amendment, as appropriate, may significantly reduce contaminant concentrations after a long time allowing for acclimation. This is typically much less expensive than excavation followed by disposal elsewhere, incineration or other *ex situ* treatment strategies, and reduces or eliminates the need for "pump and treat", a practice common at sites where hydrocarbons have contaminated clean groundwater.

Monitoring Bioremediation

The process of bioremediation can be monitored indirectly by measuring the *Oxidation Reduction Potential* or redox in soil and groundwater, together with pH, temperature, oxygen content, electron acceptor/donor concentrations, and concentration of breakdown products (e.g. carbon dioxide).

This, by itself and at a single site, gives little information about the process of remediation.

1. It is necessary to sample enough points on and around the contaminated site to be able to determine contours of equal redox potential. Contouring is usually done using specialised software, e.g. using Kriging interpolation.
2. If all the measurements of redox potential show that electron acceptors have been used up, it is in effect an indicator for total microbial activity. Chemical analysis is also required to determine when the levels of contaminants and their breakdown products have been reduced to below regulatory limits.

Microbial Biodegradation

Interest in the microbial biodegradation of pollutants has intensified in recent years as humanity strives to find sustainable ways to cleanup contaminated environments. These bioremediation and biotransformation methods endeavour to harness the astonishing, naturally occurring, ability of microbial xenobiotic metabolism to degrade, transform or accumulate a huge range of compounds including hydrocarbons (e.g. oil), polychlorinated biphenyls (PCBs), polyaromatic hydrocarbons (PAHs), heterocyclic compounds (such as pyridine or quinoline), pharmaceutical substances, radionuclides and metals. Major methodological breakthroughs in recent years have enabled detailed genomic, metagenomic, proteomic, bioinformatic and other high-throughput analyses of environmentally relevant microorganisms providing unprecedented insights into key biodegradative pathways and the ability of organisms to adapt to changing environmental conditions.

The elimination of a wide range of pollutants and wastes from the environment is an absolute requirement to promote a sustainable development of our society with low environmental impact. Biological processes play a major role in the removal of contaminants and they take advantage of the astonishing catabolic versatility of microorganisms to degrade/convert such compounds. New methodological breakthroughs in sequencing, genomics, proteomics, bioinformatics and imaging are producing vast amounts of information. In the field of Environmental Microbiology, genome-based global studies open a new era providing unprecedented *in silico* views of metabolic and regulatory networks, as well as clues to the evolution of degradation pathways and to the molecular adaptation strategies to changing environmental conditions. Functional genomic and metagenomic approaches are increasing our understanding of the relative importance of different pathways and regulatory networks to carbon flux in particular environments and for particular compounds and they will certainly accelerate the development of bioremediation technologies and biotransformation processes.

Aerobic Biodegradation of Pollutants

The burgeoning amount of bacterial genomic data provides unparalleled opportunities for understanding the genetic and molecular bases of the degradation of organic pollutants. Aromatic compounds are among the most recalcitrant of these pollutants and lessons can be learned from the recent genomic studies of *Burkholderia xenovorans* LB400 and *Rhodococcus* sp. strain RHA1, two of the largest bacterial genomes completely sequenced to date. These studies have helped expand our understanding of bacterial catabolism, non-catabolic physiological adaptation to organic compounds, and the evolution of large bacterial genomes. First, the metabolic pathways from phylogenetically diverse isolates are very similar with respect to overall organization. Thus, as originally noted in pseudomonads, a large number of "peripheral aromatic" pathways funnel a range of natural and xenobiotic compounds into a restricted number of "central aromatic" pathways. Nevertheless, these pathways are genetically organized in genus-specific fashions, as exemplified by the b-ketoadipate and Paa pathways. Comparative genomic studies further reveal that some pathways are more widespread than initially thought. Thus, the Box and Paa pathways illustrate the prevalence of non-oxygenolytic ring-cleavage strategies in aerobic aromatic degradation processes. Functional genomic studies have been useful in establishing that even organisms harboring high numbers of homologous enzymes seem to contain few examples of true redundancy. For example, the multiplicity of ring-cleaving dioxygenases in certain rhodococcal isolates may be attributed to the cryptic aromatic catabolism of different terpenoids and steroids. Finally, analyses have indicated that recent genetic flux appears to have played a more significant role in the evolution of some large genomes, such as LB400's, than others. However, the emerging trend is that the large gene repertoires of potent pollutant degraders such as LB400 and RHA1 have evolved principally through more ancient processes. That this is true in such phylogenetically diverse species is remarkable and further suggests the ancient origin of this catabolic capacity.

Anaerobic Biodegradation of Pollutants

Anaerobic microbial mineralization of recalcitrant organic pollutants is of great environmental significance and involves intriguing novel biochemical reactions. In particular, hydrocarbons and halogenated compounds have long been doubted to be degradable in the absence of oxygen, but the isolation of hitherto unknown anaerobic hydrocarbon-degrading and reductively dehalogenating bacteria during the last decades provided ultimate proof for these processes in nature. Many novel biochemical reactions were discovered enabling the respective metabolic pathways, but progress in the molecular understanding of these bacteria was rather slow, since genetic systems are not readily applicable for most of them. However, with the increasing application of genomics in the field of environmental microbiology, a new and promising perspective is now at hand to obtain molecular insights into these new metabolic properties. Several complete genome sequences were determined during the last few years from bacteria capable of anaerobic organic pollutant degradation. The ~4.7 Mb genome of the facultative denitrifying *Aromatoleum aromaticum* strain EbN1 was the first to be determined for an anaerobic hydrocarbon degrader (using toluene or ethylbenzene as substrates).

The genome sequence revealed about two dozen gene clusters (including several paralogs) coding for a complex catabolic network for anaerobic and aerobic degradation of aromatic compounds. The genome sequence forms the basis for current detailed studies on regulation of pathways and enzyme structures. Further genomes of anaerobic hydrocarbon degrading bacteria were recently completed for the iron-reducing species *Geobacter metallireducens* (accession nr. NC_007517) and the perchlorate-reducing *Dechloromonas aromatica* (accession nr. NC_007298), but these are not yet evaluated in formal publications. Complete genomes were also determined for bacteria capable of anaerobic degradation of halogenated hydrocarbons by halorespiration: the ~1.4 Mb genomes of *Dehalococcoides ethenogenes* strain 195 and *Dehalococcoides* sp. strain CBDB1 and the ~5.7 Mb genome of *Desulfitobacterium hafniense* strain Y51. Characteristic for all these bacteria is the presence of multiple paralogous genes for reductive dehalogenases, implicating a wider dehalogenating spectrum of the organisms than previously known. Moreover, genome sequences provided unprecedented insights into the evolution of reductive dehalogenation and differing strategies for niche adaptation.

Recently, it has become apparent that some organisms, including *Desulfitobacterium chlororespirans*, originally evaluated for halorespiration on chlorophenols, can also use certain brominated compounds, such as the herbicide bromoxynil and its major metabolite as electron acceptors for growth. Iodinated compounds may be dehalogenated as well, though the process may not satisfy the need for an electron acceptor.

Bioavailability, Chemotaxis, and Transport of Pollutants

Bioavailability, or the amount of a substance that is physiochemically accessible to microorganisms is a key factor in the efficient biodegradation of pollutants. O'Loughlin et al (2000) showed that, with the exception of kaolinite clay, most soil clays and cation exchange resins attenuated biodegradation of 2-picoline by *Arthrobacter* sp. strain R1, as a result of adsorption of the substrate to the clays. Chemotaxis, or the directed movement of motile organisms towards or away from chemicals in the environment is an important physiological response that may contribute to effective catabolism of molecules in the environment. In addition, mechanisms for the intracellular accumulation of aromatic molecules via various transport mechanisms are also important.

Oil Biodegradation

Petroleum oil contains aromatic compounds that are toxic for most life forms. Episodic and chronic pollution of the environment by oil causes major ecological perturbations. Marine environments are especially vulnerable since oil spills of coastal regions and the open sea are poorly containable and mitigation is difficult. In addition to pollution through human activities, about 250 million litres of petroleum enter the marine environment every year from natural seepages. Despite its toxicity, a considerable fraction of petroleum oil entering marine systems is eliminated by the hydrocarbon-degrading activities of microbial communities, in particular by a remarkable recently discovered group of specialists, the so-called hydrocarbonoclastic bacteria (HCB). *Alcanivorax borkumensis* was the first HCB to have its genome sequenced. In addition to hydrocarbons, crude oil often contains various heterocyclic compounds, such as pyridine, which appear to be degraded by similar, though separate mechanisms than hydrocarbons.

Analysis of Waste Biotreatment

Sustainable development requires the promotion of environmental management and a constant search for new technologies to treat vast quantities of wastes generated by increasing anthropogenic activities. Biotreatment, the processing of wastes using living organisms, is an environmentally friendly, relatively simple and cost-effective alternative to physico-chemical clean-up options. Confined environments, such as bioreactors, have been engineered to overcome the physical, chemical and biological limiting factors of biotreatment processes in highly controlled systems.

The great versatility in the design of confined environments allows the treatment of a wide range of wastes under optimized conditions. To perform a correct assessment, it is necessary to consider various microorganisms having a variety of genomes and expressed transcripts and proteins. A great number of analyses are often required. Using traditional genomic techniques, such assessments are limited and time-consuming. However, several high-throughput techniques originally developed for medical studies can be applied to assess biotreatment in confined environments.

Metabolic Engineering and Biocatalytic Applications

The study of the fate of persistent organic chemicals in the environment has revealed a large reservoir of enzymatic reactions with a large potential in preparative organic synthesis, which has already been exploited for a number of oxygenases on pilot and even on industrial scale. Novel catalysts can be obtained from metagenomic libraries and DNA sequence based approaches.

Our increasing capabilities in adapting the catalysts to specific reactions and process requirements by rational and random mutagenesis broadens the scope for application in the fine chemical industry, but also in the field of biodegradation. In many cases, these catalysts need to be exploited in whole cell bioconversions or in fermentations, calling for system-wide approaches to understanding strain physiology and metabolism and rational approaches to the engineering of whole cells as they are increasingly put forward in the area of systems biotechnology and synthetic biology.

Fungal Biodegradation

In the ecosystem, different substrates are attacked at different rates by consortia of organisms from different kingdoms. *Aspergillus* and other moulds play an important role in these consortia because they are adept at recycling starches, hemicelluloses, celluloses, pectins and other sugar polymers. Some aspergilli are capable of degrading more refractory compounds such as fats, oils, chitin, and keratin. Maximum decomposition occurs when there is sufficient nitrogen, phosphorus and other essential inorganic nutrients. Fungi also provide food for many soil organisms.

For *Aspergillus* the process of degradation is the means of obtaining nutrients. When these moulds degrade human-made substrates, the process usually is called biodeterioration. Both paper and textiles (cotton, jute, and linen) are particularly vulnerable to *Aspergillus* degradation. Our artistic heritage is also subject to *Aspergillus* assault. To give but one example, after Florence in Italy flooded in 1969, 74% of the isolates from a damaged Ghirlandaio fresco in the Ognissanti church were *Aspergillus versicolor*.

Mycoremediation

Mycoremediation is a form of bioremediation, the process of using fungi to degrade or sequester contaminates in the environment. Stimulating microbial and enzyme activity, mycelium reduces toxins in-situ. Some fungi are hyperaccumulators, capable of absorbing and concentrating heavy metals in the mushroom fruit bodies.

One of the primary roles of fungi in the ecosystem is decomposition, which is performed by the mycelium. The mycelium secretes extracellular enzymes and acids that break down lignin and cellulose, the two main building blocks of plant fibre. These are organic compounds composed of long chains of carbon and hydrogen, structurally similar to many organic pollutants. The key to mycoremediation is determining the right fungal species to target a specific pollutant. Certain strains have been reported to successfully degrade the nerve gases VX and sarin.

In an experiment conducted in conjunction with Dr. S. A. Thomas, a major contributor in the bioremediation industry, a plot of soil contaminated with diesel oil was inoculated with mycelia of oyster mushrooms; traditional bioremediation techniques (bacteria) were used on control plots. After four weeks, more than 95% of many of the PAH (polycyclic aromatic hydrocarbons) had been reduced to non-toxic components in the mycelial-inoculated plots. It appears that the natural microbial community participates with the fungi to break down contaminants, eventually into carbon dioxide and water. Wood-degrading fungi are particularly effective in breaking down aromatic pollutants (toxic components of petroleum), as well as chlorinated compounds (certain persistent pesticides; Battelle, 2000).

The concept of mycoremediation was explored in the 1984 film Nausicaa of the Valley of the Wind, where vast tracts of fungal forest rehabilitate the planet after catastrophic human polluting and apocalypse.

Mycofiltration is a similar or same process, using fungal mycelia to filter toxic waste and microorganisms from water in soil.

Archaea

The Archaea are a group of single-celled microorganisms. A single individual or species from this domain is called an *archaeon* (sometimes spelled "archeon"). They have no cell nucleus nor any other membrane-bound organelles within their cells. In the past they were viewed as an unusual group of bacteria and named archaebacteria, but since the Archaea have an independent evolutionary history and show many differences in their biochemistry from other forms of life, they are now classified as a separate domain in the three-domain system. In this system the phylogenetically distinct branches of evolutionary descent are the Archaea, Bacteria and Eukaryota. Archaea are divided into four recognized phyla, but many more phyla may exist. Of these groups the Crenarchaeota and the Euryarchaeota are most intensively studied. Classification is still difficult, since the vast majority have never been studied in the laboratory and have only been detected by analysis of their nucleic acids in samples from the environment. Although archaea have, in the past, been classed with bacteria as *prokaryotes* (or Kingdom Monera), this classification is regarded by some as outdated.

Archaea and bacteria are quite similar in size and shape, although a few archaea have very unusual shapes, such as the flat and square-shaped cells of *Haloquadra walsbyi*. Despite this visual similarity to bacteria, archaea possess genes and several metabolic pathways that are more closely related to those of eukaryotes: notably the enzymes involved in transcription and translation. Other aspects of archaean biochemistry are unique, such as their reliance on ether lipids in their cell membranes. The archaea exploit a much greater variety of sources of energy than eukaryotes: ranging from familiar organic compounds such as sugars, to using ammonia, metal ions or even hydrogen gas as nutrients. Salt-tolerant archaea (the Halobacteria) use sunlight as an energy source and other species of archaea fix carbon; however, unlike plants and cyanobacteria, no species of archaea is known to do both. Archaea reproduce asexually and divide by binary fission, fragmentation, or budding; in contrast to bacteria and eukaryotes, no known species form spores.

Initially, archaea were seen as extremophiles that lived in harsh environments, such as hot springs and salt lakes, but they have since been found in a broad range of habitats, including soils, oceans, and marshlands. Archaea are particularly numerous in the oceans, and the archaea in plankton may be one of the most abundant groups of organisms on the planet. Archaea are now recognized as a major part of Earth's life and may play roles in both the carbon cycle and the nitrogen cycle. No clear examples of archaeal pathogens or parasites are known, but they are often mutualists or commensals. One example are the methanogens that inhabit the gut of humans and ruminants, where their vast numbers aid digestion. Methanogens are used in biogas production and sewage treatment, and enzymes from extremophile archaea that can endure high temperatures and organic solvents are exploited in biotechnology.

Classification

A New Domain

Early in the 20th century, prokaryotes were regarded as a single group of organisms and classified based on their biochemistry, morphology and metabolism. For example, microbiologists tried to classify microorganisms based on the structures of their cell walls, their shapes, and the substances they consume. However, a new approach was proposed in 1965, using the sequences of the genes in these organisms to work out which prokaryotes are genuinely related to each other. This approach, known as phylogenetics, is the main method used today.

Archaea were first classified as a separate group of prokaryotes in 1977 by Carl Woese and George E. Fox in phylogenetic trees based on the sequences of ribosomal RNA (rRNA) genes. These two groups were originally named the Archaebacteria and Eubacteria and treated as kingdoms or subkingdoms, which Woese and Fox termed *Urkingdoms*. Woese argued that this group of prokaryotes is a fundamentally different sort of life. To emphasize this difference, these two domains were later renamed Archaea and Bacteria.

At first, only the methanogens were placed in this new domain, and the archaea were seen as extremophiles that exist only in habitats such as hot springs and salt lakes. By the end of the 20th century, microbiologists realized that archaea is a large and diverse group of organisms that are widely distributed in nature and are common in much less extreme habitats, such as soils and oceans. This new appreciation of the importance and ubiquity of archaea came from using the polymerase chain reaction to detect prokaryotes in samples of water or soil from their nucleic acids alone. This allows the detection and identification of organisms that cannot be cultured in the laboratory, which generally remains difficult.

Current Classification

The classification of archaea, and of prokaryotes in general, is a rapidly moving and contentious field. Current classification systems aim to organize archaea into groups of organisms that share structural features and common ancestors. These classifications rely heavily on the use of the sequence of ribosomal RNA genes to reveal relationships between organisms (molecular phylogenetics). Most of the culturable and well-investigated species of archaea are members of two main phyla, the Euryarchaeota and Crenarchaeota. Other groups have been tentatively created.

For example, the peculiar species *Nanoarchaeum equitans*, which was discovered in 2003, has been given its own phylum, the Nanoarchaeota. A new phylum Korarchaeota has also been proposed. It contains a small group of unusual thermophilic species that shares features of both of the main phyla, but is most closely related to the Crenarchaeota. Other recently detected species of archaea are only distantly related to any of these groups, such as the Archaeal Richmond Mine Acidophilic Nanoorganisms (ARMAN), which were discovered in 2006 and are some of the smallest organisms known.

Species?

The classification of archaea into species is also controversial. Biology defines a species as a group of related organisms. The familiar exclusive breeding criterion (organisms that can breed with each other but not with others), is of no help because archaea reproduce asexually.

Archaea show high levels of horizontal gene transfer between lineages. Some researchers suggest that individuals can be grouped into species-like populations given highly similar genomes and infrequent gene transfer to/from cells with less-related genomes, as in the genus *Ferroplasma*. On the other hand, studies in *Halorubrum* found significant genetic transfer to/from less-related populations, limiting the criteria's applicability. A second concern is to what extent such species designations have practical meaning.

Current knowledge on genetic diversity is fragmentary and the total number of archaean species cannot be estimated with any accuracy. Estimates of the number of phyla range from 18 to 23, of which only 8 have representatives that have been cultured and studied directly. Many of these hypothesized groups are known from a single rRNA sequence, indicating that the diversity among these organisms remains obscure. The Bacteria also contain many uncultured microbes with similar implications for characterization.

Origin and Evolution

Although probable prokaryotic cell fossils date to almost 3.5 billion years ago, most prokaryotes do not have distinctive morphologies and fossil shapes cannot be used to identify them as Archaea. Instead, chemical fossils of unique lipids are more informative because such compounds do not occur in other organisms. Some publications suggest that archaean or eukaryotic lipid remains are present in shales dating from 2.7 billion years ago; such data have since been questioned. Such lipids have also been detected in Precambrian formations. The oldest such traces come from the Isua district of west Greenland, which include Earth's oldest sediments, formed 3.8 billion years ago. The archaeal lineage may be the most ancient that exists on earth.Woese argued that the bacteria, archaea, and eukaryotes represent separate lines of descent that diverged early on from an ancestral colony of organisms. A few biologists, however, argue that the Archaea and Eukaryota arose from a group of bacteria. In any case it is thought that viruses and archaea began relationships approximately two billion years ago, and that co-evolution may have been occurring between members of these groups. It is possible that the last common ancestor of the bacteria and archaea was a thermophile, which raises the possibility that lower temperatures are "extreme environments" in archaeal terms, and organisms that live in cooler environments appeared only later. Since the Archaea and Bacteria are no more related to each other than they are to eukaryotes, the term *prokaryote*'s only surviving meaning is "not a eukaryote", limiting its value.

Archaea and Eukaryotes

The relationship between archaea and eukaryotes remains problematic. Aside from the similarities in cell structure and function that are discussed below, many genetic trees group the two.

Complicating factors include claims that the relationship between eukaryotes and the archaeal phylum Euryarchaeota is closer than the relationship between the Euryarchaeota and the phylum Crenarchaeota and the presence of archaean-like genes in certain bacteria, such as *Thermotoga maritima*, from horizontal gene transfer. The leading hypothesis is that the ancestor of the eukaryotes diverged early from the Archaea, and that eukaryotes arose through fusion of an archaean and eubacterium, which became the nucleus and cytoplasm; this accounts for various genetic similarities but runs into difficulties explaining cell structure.

Morphology

Individual archaea range from 0.1 micrometers (μm) to over 15 μm in diameter, and occur in various shapes, commonly as spheres, rods, spirals or plates. Other morphologies in the Crenarchaeota include irregularly shaped lobed cells in *Sulfolobus*, needle-like filaments that are less than half a micrometer in diameter in *Thermofilum*, and almost perfectly rectangular rods in *Thermoproteus* and *Pyrobaculum*. *Haloquadra walsbyi* are flat, square archaea that live in hypersaline pools. These unusual shapes are probably maintained both by their cell walls and a prokaryotic cytoskeleton. Proteins related to the cytoskeleton components of other organisms exist in archaea, and filaments form within their cells, but in contrast to other organisms, these cellular structures are poorly understood. In *Thermoplasma* and *Ferroplasma* the lack of a cell wall means that the cells have irregular shapes, and can resemble amoebae.

Some species form aggregates or filaments of cells up to 200 μm long. These organisms can be prominent in biofilms. Notably, aggregates of *Thermococcus coalescens* cells fuse together in culture, forming single giant cells.

Archaea in the genus *Pyrodictium* produce an elaborate multicell colony involving arrays of long, thin hollow tubes called *cannulae* that stick out from the cells' surfaces and connect them into a dense bush-like agglomeration.

The function of these cannulae is not settled, but they may allow communication or nutrient exchange with neighbours. Multi-species colonies exist, such as the "string-of-pearls" community that was discovered in 2001 in a German swamp. Round whitish colonies of a novel Euryarchaeota species are spaced along thin filaments that can range up to 15 centimetres (5.9 in) long; these filaments are made of a particular bacteria species.

Structure, Composition Development, Operation

Archaea and bacteria have generally similar cell structure, but cell composition and organization set the archaea apart. Like bacteria, archaea lack interior membranes and organelles. Like bacteria, archaea cell membranes are usually bounded by a cell wall and they swim using one or more flagella. Structurally, archaea are most similar to gram-positive bacteria. Most have a single plasma membrane and cell wall, and lack a periplasmic space; the exception to this general rule is *Ignicoccus*, which possess a particularly large periplasm that contains membrane-bound vesicles and is enclosed by an outer membrane.

Membranes

Archaeal membranes are made of molecules that differ strongly from those in other life forms, showing that archaea are related only distantly to bacteria and eukaryotes. In all organisms cell membranes are made of molecules known as phospholipids. These molecules possess both a polar part that dissolves in water (the phosphate "head"), and a "greasy" non-polar part that does not (the lipid tail). These dissimilar parts are connected by a glycerol group. In water, phospholipids cluster, with the heads facing the water and the tails facing away from it. The major structure in cell membranes is a double layer of these phospholipids, which is called a lipid bilayer.

These phospholipids are unusual in four ways:

- Bacteria and eukaryotes have membranes composed mainly of glycerol-ester lipids, whereas archaea have membranes composed of glycerol-ether lipids. The difference is the type of bond that joins the lipids to the glycerol group; the two types are shown in yellow in the Figure at the right. In ester lipids this is an ester bond, whereas in ether lipids this is an ether bond. Ether bonds are chemically more resistant than ester bonds. This stability might help archaea to survive extreme temperatures and very acidic or alkaline environments. Bacteria and eukaryotes do contain some ether lipids, but in contrast to archaea these lipids are not a major part of their membranes.
- The stereochemistry of the glycerol group is the reverse of that found in other organisms. The glycerol group can occur in two forms that are mirror images of one another, called the right-

handed and left-handed forms; in chemistry these are called *enantiomers*. Just as a right hand does not fit easily into a left-handed glove, a right-handed glycerol molecule generally cannot be used or made by enzymes adapted for the left-handed form. This suggests that archaea use entirely different enzymes for synthesizing phospholipids than do bacteria and eukaryotes. Such enzymes developed very early in life's history, suggesting an early split from the other two domains.

- Archaeal lipid tails are chemically different from other organisms. Archaeal lipids are based upon the isoprenoid sidechain and are long chains with multiple side-branches and sometimes even cyclopropane or cyclohexane rings. This is in contrast to the fatty acids found in other organisms' membranes, which have straight chains with no branches or rings. Although isoprenoids play an important role in the biochemistry of many organisms, only the archaea use them to make phospholipids. These branched chains may help prevent archaean membranes from leaking at high temperatures.
- In some archaea the lipid bilayer is replaced by a monolayer. In effect, the archaea fuse the tails of two independent phospholipid molecules into a single molecule with two polar heads; this fusion may make their membranes more rigid and better able to resist harsh environments. For example, the lipids in *Ferroplasma* are of this type, which is thought to aid this organism's survival in its highly acidic habitat.

Wall and Flagella

Most archaea (not *Thermoplasma* and *Ferroplasma*) possess a cell wall. In most archaea the wall is assembled from surface-layer proteins, which form an S-layer. An S-layer is a rigid array of protein molecules that cover the outside of the cell (like chain mail). This layer provides both chemical and physical protection, and can prevent macromolecules from contacting the cell membrane. Unlike bacteria, most archaea lack peptidoglycan in their cell walls. The exception is pseudopeptidoglycan, which is found in Methanobacteriales, but pseudopeptidoglycan lacks D-amino acids and N-acetylmuramic acid.

Archaea flagella operate like bacterial flagella—their long stalks are driven by rotatory motors at the base. These motors are powered by the proton gradient across the membrane. However, archaeal flagella are notably different in composition and development. The two types of flagella evolved from different ancestors. The bacterial flagellum shares a common ancestor with the type III secretion system, while archaeal flagella appear to have evolved from bacterial type IV pili. In contrast to the bacterial flagellum, which is hollow and is assembled by subunits moving up the central pore to the tip of the flagella, archaeal flagella are synthesized by adding subunits at the base.

Metabolism

Archaea exhibit a great variety of chemical reactions in their metabolism and use many sources of energy. These reactions are classified into nutritional groups, depending on energy and carbon sources. Some archaea obtain energy from inorganic compounds such as sulfur or ammonia (they are lithotrophs). These include nitrifiers, methanogens and anaerobic methane oxidisers. In these reactions one compound passes electrons to another (in a redox reaction), releasing energy to fuel the cell's activities. One compound acts as an electron donor and one as an electron acceptor. The energy released generates adenosine triphosphate (ATP) through chemiosmosis, in the same basic process that happens in the mitochondrion of eukaryotic cells.

Other groups of archaea use sunlight as a source of energy (they are phototrophs). However, oxygen–generating photosynthesis does not occur in any of these organisms. Many basic metabolic pathways are shared between all forms of life; for example, archaea use a modified form of glycolysis (the Entner–Doudoroff pathway) and either a complete or partial citric acid cycle. These similarities to other organisms probably reflect both early origins in the history of life and their high level of efficiency.

Other archaea use CO_2 in the atmosphere as a source of carbon, in a process called carbon fixation (they are autotrophs). This process involves either a highly modified form of the Calvin cycle or a recently discovered metabolic pathway called the 3-hydroxypropionate/4-hydroxybutyrate cycle. The Crenarchaeota also use the reverse Krebs cycle while the Euryarchaeota also use the reductive acetyl-CoA pathway. Carbon–fixation is powered by inorganic energy sources. No known archaea carry out photosynthesis. Archaeal energy sources are extremely diverse, and range from the oxidation of ammonia by the Nitrosopumilales to the oxidation of hydrogen sulfide or elemental sulfur by species of *Sulfolobus*, using either oxygen or metal ions as electron acceptors.

Phototrophic archaea use light to produce chemical energy in the form of ATP. In the Halobacteria, light-activated ion pumps like bacteriorhodopsin and halorhodopsin generate ion gradients by pumping ions out of the cell across the plasma membrane. The energy stored in these electrochemical gradients is then converted into ATP by ATP synthase. This process is a form of photophosphorylation. The ability of these light-driven pumps to move ions across membranes depends on light-driven changes in the structure of a retinol cofactor buried in the centre of the protein.

Genetics

Archaea usually have a single circular chromosome, the size of which may be as great as 5,751,492 base pairs in *Methanosarcina acetivorans*, the largest known archaean genome. One-tenth of this size is the tiny 490,885 base-pair genome of *Nanoarchaeum equitans*, the smallest archaean genome known; it is estimated to contain only 537 protein-encoding genes. Smaller independent pieces of DNA, called *plasmids*, are also found in archaea. Plasmids may be transferred between cells by physical contact, in a process that may be similar to bacterial conjugation.

Archaea can be infected by double-stranded DNA viruses that are unrelated to any other form of virus and have a variety of unusual shapes, including bottles, hooked rods, or teardrops. These viruses have been studied in most detail in thermophilics, particularly the orders Sulfolobales and Thermoproteales. A single-stranded DNA virus that infects halophilics was identified in 2009. Defences against these viruses may involve RNA interference from repetitive DNA sequences that are related to the genes of the viruses.

Archaea are genetically distinct from bacteria and eukaryotes, with up to 15% of the proteins encoded by any one archaeal genome being unique to the domain, although most of these unique genes have no known function. Of the remainder of the unique proteins that have an identified function, most are involved in methanogenesis. The proteins that archaea, bacteria and eukaryotes share form a common core of cell function, relating mostly to transcription, translation, and nucleotide metabolism. Other characteristic archaean features are the organization of genes of related function—such as enzymes that catalyse steps in the same metabolic pathway into novel operons, and large differences in tRNA genes and their aminoacyl tRNA synthetases.

Transcription and translation in archaea resemble these processes in eukaryotes more than in bacteria, with the archaean RNA polymerase and ribosomes being very close to their equivalents in eukaryotes. Although archaea only have one type of RNA polymerase, its structure and function in transcription seems to be close to that of the eukaryotic RNA polymerase II, with similar protein assemblies (the general transcription factors) directing the binding of the RNA polymerase to a gene's promoter. However, other archaean transcription factors are closer to those found in bacteria. Post-transcriptional modification is simpler than in eukaryotes, since most archaean genes lack introns, although there are many introns in their transfer RNA and ribosomal RNA genes, and introns may occur in a few protein-encoding genes.

Reproduction

Archaea reproduce asexually by binary or multiple fission, fragmentation, or budding; meiosis does not occur, so if a species of archaea exists in more than one form, all have the same genetic material. Cell division is controlled in a cell cycle; after the cell's chromosome is replicated and the two daughter chromosomes separate, the cell divides. Details have only been investigated in the genus *Sulfolobus*, but here that cycle has characteristics that are similar to both bacterial and eukaryotic systems. The chromosomes replicate from multiple starting-points (origins of replication) using DNA polymerases that resemble the equivalent eukaryotic enzymes. However, the proteins that direct cell division, such as the protein FtsZ, which forms a contracting ring around the cell, and the components of the septum that is constructed across the centre of the cell, are similar to their bacterial equivalents.

Both bacteria and eukaryotes, but not archaea, make spores. Some species of Haloarchaea undergo phenotypic switching and grow as several different cell types, including thick-walled structures that are resistant to osmotic shock and allow the archaea to survive in water at low salt concentrations, but these are not reproductive structures and may instead help them reach new habitats.

Ecology

Habitats

Archaea exist in a broad range of habitats, and as a major part of global ecosystems, may contribute up to 20% of earth's biomass. The first-discovered archaeans were extremophiles. Indeed, some archaea survive high temperatures, often above 100 °C (212 °F), as found in geysers, black smokers, and oil wells. Other common habitats include very cold habitats and highly saline, acidic, or alkaline water. However, archaea include mesophiles that grow in mild conditions, in marshland, sewage, the oceans, and soils. Extremophile archaea are members of four main physiological groups. These are the halophiles, thermophiles, alkaliphiles, and acidophiles. These groups are not comprehensive or phylum-specific, nor are they mutually exclusive, since some archaea belong to several groups. Nonetheless, they are a useful starting point for classification.

Halophiles, including the genus *Halobacterium*, live in extremely saline environments such as salt lakes and outnumber their bacterial counterparts at salinities greater than 20–25%. Thermophiles grow best at temperatures above 45 °C (113 °F), in places such as hot springs; *hyperthermophilic* archaea grow optimally at temperatures greater than 80 °C (176 °F). The archaeal *Methanopyrus kandleri* Strain 116 grows at 122 °C (252 °F), the highest recorded temperature of any organism.

Other archaea exist in very acidic or alkaline conditions. For example, one of the most extreme archaean acidophiles is *Picrophilus torridus*, which grows at pH 0, which is equivalent to thriving in 1.2 Molar sulfuric acid. This resistance to extreme environments has made archaea the focus of speculation about the possible properties of extraterrestrial life. Extremophile habitats are not dissimilar to those on Mars, leading to the suggestion that viable microbes could be transferred between planets in meteorites.

Recently, several studies have shown that archaea exist not only in mesophilic and thermophilic environments but are also present, sometimes in high numbers, at low temperatures as well. For example, archaea are common in cold oceanic environments such as polar seas. Even more significant are the large numbers of archaea found throughout the world's oceans in non-extreme habitats among the plankton community (as part of the picoplankton). Although these archaea can be present in extremely high numbers (up to 40% of the microbial biomass), almost none of these species have been isolated and studied in pure culture. Consequently, our understanding of the role of archaea in ocean ecology is rudimentary, so their full influence on global biogeochemical cycles remains largely unexplored. Some marine Crenarchaeota are capable of nitrification, suggesting these organisms may affect the oceanic nitrogen cycle, although these oceanic Crenarchaeota may also use other sources of energy. Vast numbers of archaea are also found in the sediments that cover the sea floor, with these organisms making up the majority of living cells at depths over 1 meter into this sediment.

Role in Chemical Cycling

Archaea recycle elements such as carbon, nitrogen and sulfur through their various habitats. Although these activities are vital for normal ecosystem function, archaea can also contribute to human-made changes, and even cause pollution.

Archaea carry out many steps in the nitrogen cycle. This includes both reactions that remove nitrogen from ecosystems, such as nitrate-based respiration and denitrification, as well as processes that introduce nitrogen, such as nitrate assimilation and nitrogen fixation. Archaean involvement in ammonia oxidation reactions was recently discovered. These reactions are particularly important in the oceans. The archaea also appear to be crucial for ammonia oxidation in soils. The produce nitrite, which other microbes then oxidize to nitrate. Plants and other organisms consume the latter. In the sulfur cycle, archaea that grow by oxidizing sulfur compounds release this element from rocks, making it available to other organisms. However, the archaea that do this, such as *Sulfolobus*, produce sulfuric acid as a waste product, and the growth of these organisms in abandoned mines can contribute to acid mine drainage and other environmental damage.

In the carbon cycle, methanogen archaea remove hydrogen and are important in the decay of organic matter by the populations of microorganisms that act as decomposers in anaerobic ecosystems, such as sediments, marshes and sewage treatment works. However, methane is one of the most abundant greenhouse gases in Earth's atmosphere, constituting 18% of the global total. It is 25 times more potent as a greenhouse gas than carbon dioxide. Methanogens are the primary source of atmospheric methane, and are responsible for most of the world's yearly methane emissions. As a consequence, these archaea contribute to global greenhouse gas emissions and global warming.

Interactions with other Organisms

The well-characterized interactions between archaea and other organisms are either mutual or commensal. As of 2007, no clear examples of archaeal pathogens or parasites were known. However, a relationship has been proposed between some species of methanogens and infections in the mouth, and *Nanoarchaeum equitans* may be a parasite of another species of archaea, since it only survives and reproduces within the cells of the Crenarchaeon *Ignicoccus hospitalis*, and appears to offer no benefit to its host.

Mutualism

One well-understood example of mutualism is the interaction between protozoa and methanogenic archaea in the digestive tracts of animals that digest cellulose, such as ruminants and termites. In these anaerobic environments, protozoa break down plant cellulose to obtain energy. This process releases hydrogen as a waste product, but high levels of hydrogen reduce energy production. When methanogens convert hydrogen to methane, protozoa benefit from more energy. In anaerobic protozoa such as *Plagiopyla frontata* the archaea reside inside the protozoa and consume hydrogen produced in their hydrogenosomes. Archaea associate with larger organisms. For example, the marine archaean *Cenarchaeum symbiosum* lives within (it is an endosymbiont of) the sponge *Axinella mexicana*.

Commensalism

Archaea can also be commensals, benefiting from an association without helping or harming the other organism. For example, the methanogen *Methanobrevibacter smithii* is by far the most common archaean in the human flora, making up about one in ten of all the prokaryotes in the human gut. In termites and in humans, these methanogens may in fact be mutualists, interacting with other microbes in the gut to aid digestion. Archaean communities also associate with a range of other organisms, such as on the surface of corals, and in the region of soil that surrounds plant roots (the rhizosphere).

Significance in Technology and Industry

Extremophile archaea, particularly those resistant either to heat or to extremes of acidity and alkalinity, are a source of enzymes that function under these harsh conditions. These enzymes have found many uses.

For example, thermostable DNA polymerases, such as the Pfu DNA polymerase from *Pyrococcus furiosus*, revolutionized molecular biology by allowing the polymerase chain reaction to be used in research as a simple and rapid technique for cloning DNA.

In industry, amylases, galactosidases and pullulanases in other species of *Pyrococcus* that function at over 100 °C (212 °F) allow food processing at high temperatures, such as the production of low lactose milk and whey. Enzymes from these thermophilic archaea also tend to be very stable in organic solvents, allowing their use in environmentally friendly processes in green chemistry that synthesize organic compounds. This stability makes them easier to use in structural biology. Consequently the counterparts of bacterial or eukaryotic enzymes from extremophile archaea are often used in structural studies.

In contrast to the range of applications of archaean enzymes, the use of the organisms themselves in biotechnology is less developed. Methanogenic archaea are a vital part of sewage treatment, since they are part of the community of microorganisms that carry out anaerobic digestion and produce biogas. In mineral processing, acidophilic archaea display promise for the extraction of metals from ores, including gold, cobalt and copper.

Archaea host a new class of potentially useful antibiotics. A few of these archaeocins have been characterized, but hundreds more are believed to exist, especially within *Haloarchaea* and *Sulfolobus*. These compounds have different structure than bacterial antibiotics, so they may have novel modes of action. In addition, they may allow the creation of new selectable markers for use in archaeal molecular biology.

Fluorescent in Situ Hybridization

Fish (fluorescence *in situ* hybridization) is a cytogenetic technique developed by Christoph Lengauer that is used to detect and localize the presence or absence of specific DNA sequences on chromosomes. FISH uses fluorescent probes that bind to only those parts of the chromosome with which they show a high degree of sequence similarity. Fluorescence microscopy can be used to find out where the fluorescent probe bound to the chromosomes. FISH is often used for finding specific features in DNA for use in genetic counselling, medicine, and species identification. FISH can also be used to detect and localize specific mRNAs within tissue samples. In this context, it can help define the spatial-temporal patterns of gene expression within cells and tissues.

Probes

Probes are often derived from fragments of DNA that were isolated, purified, and amplified for use in the Human Genome Project. The size of the human genome is so large, compared to the length that could be sequenced directly, that it was necessary to divide the genome into fragments. (In the eventual analysis, these fragments were put into order by digesting a copy of each fragment into still smaller fragments using sequence-specific endonucleases, measuring the size of each small fragment using size-exclusion chromatography, and using that information to determine where the large fragments overlapped one another.)

To preserve the fragments with their individual DNA sequences, the fragments were added into a system of continually replicating bacteria populations. Clonal populations of bacteria, each population maintaining a single artificial chromosome, are stored in various laboratories around the world. The artificial chromosomes (BAC) can be grown, extracted, and labelled, in any lab. These fragments are on the order of 100 thousand base-pairs, and are the basis for most FISH probes.

Preparation and Hybridization Process

First, a probe is constructed. The probe must be large enough to hybridize specifically with its target but not so large as to impede the hybridization process. The probe is tagged directly with fluorophores, with targets for antibodies or with biotin. Tagging can be done in various ways, such as nick translation, or PCR using tagged nucleotides.

Then, an interphase or metaphase chromosome preparation is produced. The chromosomes are firmly attached to a substrate, usually glass.

Repetitive DNA sequences must be blocked by adding short fragments of DNA to the sample. The probe is then applied to the chromosome DNA and incubated for approximately 12 hours while hybridizing. Several wash steps remove all unhybridized or partially-hybridized probes. The results are then visualized and quantified using a microscope that is capable of exciting the dye and recording images.

If the fluorescent signal is weak, amplification of the signal may be necessary in order to exceed the detection threshold of the microscope. Fluorescent signal strength depends on many factors such as probe labeling efficiency, the type of probe, and the type of dye. Fluorescently-tagged antibodies or streptavidin are bound to the dye molecule. These secondary components are selected so that they have a strong signal.

Fish experiments designed to detect or localize gene expression within cells and tissues rely on the use of a reporter gene, such as one expressing green fluorescent protein, to provide the fluorescence signal.

Variations on Probes and Analysis

Fish is a very general technique. The differences between the various FISH techniques are usually due to variations in the sequence and labeling of the probes; and how they are used in combination. These few modifications make possible all FISH techniques.

Probe size is important because longer probes hybridize more specifically than shorter probes. The overlap defines the resolution of detectable features. For example, if the goal of an experiment is to detect the breakpoint of a translocation, then the overlap of the probes — the degree to which one DNA sequence is contained in the adjacent probes — defines the minimum window in which the breakpoint may be detected.

The mixture of probe sequences determines the type of feature the probe can detect. Probes that hybridize along an entire chromosome are used to count the number of a certain chromosome, show translocations, or identify extra-chromosomal fragments of chromatin. This is often called "whole-chromosome painting." If every possible probe is used, every chromosome, (the whole genome) would be marked fluorescently, which would not be particularly useful for determining features of individual sequences.

However, a mixture of smaller probes can be created that is specific to a particular region (locus) of DNA; these mixtures are used to detect deletion mutations. When combined with a specific colour, a locus-specific probe mixture is used to detect very specific translocations. Special locus-specific probe mixtures are often used to count chromosomes, by binding to the centromeric regions of chromosomes, which are unique enough to identify each chromosome (with the exception of Chromosome 13, 14 21, 22.)

A variety of other techniques use mixtures of differently-coloured probes. A range of colours in mixtures of fluorescent dyes can be detected, so each human chromosome can be identified by a characteristic colour using whole-chromosome probe mixtures and a variety of ratios of colours.

Although there are more chromosomes than easily-distinguishable fluorescent dye colours, ratios of probe mixtures can be used to create *secondary* colours. Similar to comparative genomic hybridization, the probe mixture for the secondary colours is created by mixing the correct ratio of two sets of differently-coloured probes for the same chromosome.

This technique is sometimes called M-FISH. The same physics that make a variety of colours possible for M-FISH can be used for the detection of translocations. That is, colours that are adjacent appear to overlap; a secondary colour is observed. Some assays are designed so that the secondary colour will be present or absent in cases of interest.

An example is the detection of BCR/ABL translocations, where the secondary colour indicates disease. This variation is often called double-fusion FISH or D-FISH. In the opposite situation—where the absence of the secondary colour is pathological—is illustrated by an assay used to investigate translocations where only one of the breakpoints is known or constant.

Locus-specific probes are made for one side of the breakpoint and the other intact chromosome. In normal cells, the secondary colour is observed, but only the primary colour is observed when the translocation occurs. This technique is sometimes called "break-apart FISH".

Single Molecule RNA Fish

Single Molecule RNA Fish is a method of detecting and quantifying mRNA and other long RNA molecules in a thin layer of tissue sample. Targets can be reliably imaged through the application of multiple short singly labelled oligonucleotide probes.

The probes cooperatively bind to the target site. When each probe binds to the single stranded mRNA, it causes cooperative unwinding of the mRNA, promoting the binding of the next probe.

The net result is the binding of 48 fluorescent labels to a single molecule of mRNA, providing sufficient fluorescence to reliably locate each target mRNA in a wide-field fluorescent microscopy image. Probes not binding to the intended sequence do not achieve sufficient localized fluorescence to be distinguished from the background. This technology is exclusively licensed to Biosearch Technologies as Stellaris™ FISH Probes.

Single molecule RNA FISH assays can be performed in simplex or multiplex, and have potential applications in cancer diagnosis , neuroscience , gene expression analysis, and companion diagnostics.

Fibre Fish

In an alternative technique to interphase or metaphase preparations, fibre FISH, interphase chromosomes are attached to a slide in such a way that they are stretched out in a straight line, rather than being tightly coiled, as in conventional FISH, or adopting a random conformation, as in interphase FISH. This is accomplished by applying mechanical shear along the length of the slide, either to cells that have been fixed to the slide and then lysed, or to a solution of purified DNA.

A technique known as chromosome combing is increasingly used for this purpose. The extended conformation of the chromosomes allows dramatically higher resolution - even down to a few kilobases. The preparation of fibre FISH samples, although conceptually simple, is a rather skilled art, and only specialized laboratories use the technique routinely.

Q-Fish

Q-Fish combines Fish with PNAs and computer software to quantify fluorescence intensity. This technique is used routinely in telomere length research.

Flow-Fish

Flow-Fish uses flow cytometry to perform Fish automatically using per-cell fluorescence measurements.

Medical Applications

Often parents of children with a developmental delay want to know more about their child's conditions before choosing to have another child. These concerns can be addressed by analysis of the parents' and child's DNA. In cases where the child's developmental delay is not understood, the cause of it can be determined using FISH and cytogenetic techniques.

Examples of diseases that are diagnosed using FISH include Prader-Willi syndrome, Angelman syndrome, 22q13 deletion syndrome, chronic myelogenous leukemia, acute lymphoblastic leukemia, Cri-du-chat, Velocardiofacial syndrome, and Down syndrome. FISH on sperm cells is indicated for men with an abnormal somatic or meiotic karyotype as well as those with oligozoospermia, since approximately 50% of oligozoospermic men have an increased rate of sperm chromosome abnormalities. The analysis of chromosomes 21, X, and Y is enough to identify oligozoospermic individuals at risk.

In medicine, FISH can be used to form a diagnosis, to evaluate prognosis, or to evaluate remission of a disease, such as cancer. Treatment can then be specifically tailored. A traditional exam involving metaphase chromosome analysis is often unable to identify features that distinguish one disease from another, due to subtle chromosomal features; FISH can elucidate these differences. FISH can also be used to detect diseased cells more easily than standard Cytogenetic methods, which require dividing cells and requires labour and time-intensive manual preparation and analysis of the slides by a technologist. FISH, on the other hand, does not require living cells and can be quantified automatically, a computer counts the fluorescent dots present. However, a trained technologist is required to distinguish subtle differences in banding patterns on bent and twisted metaphase chromosomes.

Species Identification

Fish is often used in clinical studies. If a patient is infected with a suspected pathogen, bacteria, from the patient's tissues or fluids, are typically grown on agar to determine the identity of the pathogen. Many bacteria, however, even well-known species, do not grow well under laboratory conditions. FISH can be used to detect directly the presence of the suspect on small samples of patient's tissue.

Fish can also be used to compare the genomes of two biological species, to deduce evolutionary relationships. A similar hybridization technique is called a zoo blot. Bacterial FISH probes are often primers for the 16s rRNA region.

Fish is widely used in the field of microbial ecology, to identify microorganisms. Biofilms, for example, are composed of complex (often) multi-species bacterial organizations. Preparing DNA probes for one species and performing Fish with this probe allows one to visualize the distribution of this specific species within the biofilm. Preparing probes (in two different colours) for two species allows to visualize/study co-localization of these two species in the biofilm, and can be useful in determining the fine architecture of the biofilm.

Chapter 3

Microbodies

A microbody is a cytoplasmic organelle of a more or less globular shape that comprises degradative enzymes bound within a single membrane. Microbodies are specialized as containers for metabolic activity. Types include peroxisomes, glyoxisomes, glycosomes and Woronin bodies. Peroxisomes contain enzymes of b-oxidation (break down fats and produce Acetyl-CoA), as well as enzymes of many other important pathways like amino acid and bile acid metabolism, oxidation/ detoxification of various harmful compounds in the liver (ex. alcohol).

Glyoxysomes are found in germinating seeds of plants as well as in filamentous fungi. Glyoxysomes are peroxisomes with additional function - glyoxylate cycle. Glycosomes, besides peroxisomal enzymes, also possess glycolysis enzymes and are found in kinetoplastida like Trypanosomes. Woronin bodies are special organelles found only in filamentous fungi. One established function of Woronin bodies is the plugging of the septal pores after hyphal wounding, which restricts the loss of cytoplasm to the sites of injury.

Peroxisomes are organelles present in almost all eukaryotic cells. They participate in the metabolism of fatty acids and many other metabolites. Peroxisomes harbour enzymes that rid the cell of toxic peroxides. Peroxisomes are bound by a single membrane that separates their contents from the cytosol (the internal fluid of the cell) and contain membrane proteins critical for various functions, such as importing proteins into the organelles and aiding in proliferation. Peroxisomes are formed from the endoplasmic reticulum. Peroxisomes were identified as organelles by the Belgian cytologist Christian de Duve in 1967 after they had been first described in a PhD thesis of Rhodin a decade earlier.

Occurrence and Evolution

Peroxisomes are found in virtually all eukaryotic cells. Peroxisomes contain enzymes for certain oxidative reactions, like the beta-oxidation of very-long-chain fatty acids.

Prokaryotes lack peroxisomes. The enzymatic content of peroxisomes varies across species, but the presence of certain proteins common to many species has been used to suggest an endosymbiotic origin; that is, peroxisomes evolved from bacteria that invaded larger cells as parasites, and very gradually evolved a symbiotic relationship. However, this view has been challenged by recent discoveries. For example, peroxisome-less mutants can restore peroxisomes upon introduction of the wild-type gene.

Two independent evolutionary analyses of the peroxisomal proteome found homologies between the peroxisomal import machinery and the ERAD pathway in the endoplasmic reticulum, along with a number of metabolic enzymes that were likely recruited from the mitochondria. These results indicate that the peroxisome does not have an endosymbiotic origin; instead, it likely originates from the ER, and its proteins were recruited from pools existing within the primitive eukaryote, as quoted in the science textbook Biozone. However, latest research have suggested that the peroxisome may have had an actinobacterial origin, although this is still controversial.

Other organelles of the Microbody family related to peroxisomes include glyoxysomes of plants and filamentous fungi, glycosomes of kinetoplastids and Woronin bodies of filamentous fungi.

Function

Peroxisomes contain oxidative enzymes, such as catalase, D-amino acid oxidase, and uric acid oxidase. However the last enzyme is absent in humans, explaining the disease known as gout, caused by the accumulation of uric acid. Certain enzymes within the peroxisome, by using molecular oxygen, remove hydrogen atoms from specific organic substrates (labeled as R), in an oxidative reaction, producing hydrogen peroxide (H_2O_2, itself toxic):

$$RH_2 + O_2 \rightarrow R + H_2O_2$$

Catalase, another enzyme in the peroxisome, in turn uses this H_2O_2 to oxidize other substrates, including phenols, formic acid, formaldehyde, and alcohol, by means of the peroxidation reaction:

$H_2O_2 + R'H_2 \rightarrow R' + 2H_2O$, thus eliminating the poisonous hydrogen peroxide in the process.

This reaction is important in liver and kidney cells, where the peroxisomes detoxify various toxic substances that enter the blood. About 25% of the ethanol we drink is oxidized to acetaldehyde in this way. In addition, when excess H_2O_2 accumulates in the cell, catalase converts it to H_2O through this reaction:

$$2H_2O_2 \rightarrow 2H_2O + O_2$$

A major function of the peroxisome is the breakdown of fatty acid molecules, in a process called beta-oxidation. In this process, the fatty acids are broken down two carbons at a time, converted to Acetyl-CoA, which is then transported back to the cytosol for further use. In animal cells, beta-oxidation can also occur in the mitochondria. In yeast and plant cells, this process is exclusive for the peroxisome.

The first reactions in the formation of plasmalogen in animal cells also occur in peroxisomes. Plasmalogen is the most abundant phospholipid in myelin. Deficiency of plasmalogens causes profound abnormalities in the myelination of nerve cells, which is one of the reasons that many peroxisomal disorders lead to neurological disease (adrenoleukodystrophy). Peroxisomes also play a role in the production of bile acids and proteins.

In higher plants, peroxisomes contain also a complex battery of antioxidative enzymes such as superoxide dismutase, the components of the ascorbate-glutathione cycle, and the NADP-dehydrogenases of the pentose-phosphate pathway. It has been demonstrated the generation of superoxide (O_2) and nitric oxide ($^{\cdot}NO$) radicals.

The peroxisome of plant cells is polarised when fighting fungal penetration. Infection causes a glucosinolate molecule to play an antifungal role to be made and delivered to the outside of the cell through the action of the peroxisomal proteins (PEN2 and PEN3).

Protein Import

Proteins are selectively imported into peroxisomes. Since the organelles contain no DNA or ribosomes and, thus, have no means of producing proteins, all of their proteins must be imported across the membrane. It is believed that necessary proteins enter through the endoplasmic reticulum during biogenesis as well as through membrane proteins.

A specific protein signal (PTS or peroxisomal targeting signal) of three amino acids at the *C-terminus* of many peroxisomal proteins signals the membrane of the peroxisome to import them into the organelle. Other peroxisomal proteins contain a signal at the *N-*

terminus. There are at least 32 known peroxisomal proteins, called peroxins, which participate in the process of importing proteins by means of *ATP hydrolysis*. Proteins do not have to unfold to be imported into the peroxisome. The protein receptors, the peroxins *PEX5* and *PEX7*, accompany their cargoes (containing a PTS1 or a PTS2, respectively) all the way into the peroxisome where they release the cargo and then return to the cytosol - a step named *recycling*. Overall, the import cycle is referred to as the *extended shuttle mechanism*. Evidence now indicates that ATP hydrolysis is required for the recycling of receptors to the cytosol. Also, ubiquitination appears to be crucial for the export of PEX5 from the peroxisome, to the cytosol. Little is known about the import of PEX7, although it has helper proteins that have been shown to be ubiquitinated. It is interesting to note that Pex7 helper proteins share many similarities with Pex5, and, in higher eukaryotes, they are functionally replaced by a long isoform of PEX5 (PEX5L).

Deficiencies

Peroxisomal disorders are a class of conditions that lead to disorders of lipid metabolism and diseases of the nervous system. Well-known examples are X-linked adrenoleukodystrophy, the most frequent, and Zellweger syndrome. Peroxisomes matrix proteins are synthesized on free ribosomes in the cytosol and are imported post-translationally in pre-existing vesicles.

Genes

Genes that encode peroxisomal proteins include:

- PEX1
- PEX2 - PXMP3
- PEX3
- PEX5
- PEX6
- PEX7
- PEX10
- PEX11A, PEX11B, PEX11G
- PEX12
- PEX13
- PEX14
- PEX16
- PEX19

- PEX26
- PEX28
- PEX30
- PEX31

Glyoxysome

Glyoxysomes are specialized peroxisomes found in plants (particularly in the fat storage tissues of germinating seeds) and also in filamentous fungi. As in all peroxisomes, in glyoxysomes the fatty acids are hydrolyzed to acetyl-CoA by peroxisomal α-oxidation enzymes. Besides peroxisomal functions, glyoxysomes possess additionally the key enzymes of glyoxylate cycle (isocitrate lyase and malate synthase) which accomplish the glyoxylate cycle bypass.

Thus, glyoxysomes (as all peroxisomes) contain enzymes that initiate the breakdown of fatty acids and additionally possess the enzymes to produce intermediate products for the synthesis of sugars by gluconeogenesis. The seedling uses these sugars synthesized from fats until it is mature enough to produce them by photosynthesis.

Ribosomes

Ribosomes are the components of cells that make proteins from all amino acids. One of the central tenets of biology, often referred to as the "central dogma," is that DNA is used to make RNA, which, in turn, is used to make protein. The DNA sequence in genes is copied into a messenger RNA (mRNA). Ribosomes then read the information in this RNA and use it to create proteins. This process is known as translation; i.e., the ribosome "translates" the genetic information from RNA into proteins. Ribosomes do this by binding to an mRNA and using it as a template for the correct sequence of amino acids in a particular protein. The amino acids are attached to transfer RNA (tRNA) molecules, which enter one part of the ribosome and bind to the messenger RNA sequence. The attached amino acids are then joined together by another part of the ribosome. The ribosome moves along the mRNA, "reading" its sequence and producing a chain of amino acids. Ribosomes are made from complexes of RNAs and proteins. Ribosomes are divided into two subunits, one larger than the other. The smaller subunit binds to the mRNA, while the larger subunit binds to the tRNA and the amino acids. When a ribosome finishes reading a mRNA, these two subunits split apart. Ribosomes have been classified as ribozymes, since the ribosomal RNA seems to be most important for the peptidyl transferase activity that links amino acids together.

Ribosomes from bacteria, archaea and eukaryotes (the three domains of life on Earth), have significantly different structures and RNA sequences. These differences in structure allow some antibiotics to kill bacteria by inhibiting their ribosomes, while leaving human ribosomes unaffected. The ribosomes in the mitochondria of eukaryotic cells resemble those in bacteria, reflecting the likely evolutionary origin of this organelle. The word ribosome comes from *ribo*nucleic acid and the Greek: *soma* (meaning body).

Description

Archaeal, eubacterial and eukaryotic ribosomes differ in their size, composition and the ratio of protein to RNA. Because they are formed from two subunits of non-equal size, they are slightly longer in the axis than in diameter. Prokaryotic ribosomes are around 20 nm (200 ångströms) in diameter and are composed of 65% ribosomal RNA and 35% ribosomal proteins (known as a ribonucleoprotein or RNP). Eukaryotic ribosomes are between 25 and 30 nm (250-300 ångströms) in diameter and the ratio of rRNA to protein is close to 1. Ribosomes translate messenger RNA (mRNA) and build polypeptide chains (e.g., proteins) using amino acids delivered by transfer RNA (tRNA). Their active sites are made of RNA, so ribosomes are now classified as "ribozymes".

Ribosomes build proteins from the genetic instructions held within messenger RNA. Free ribosomes are suspended in the cytosol (the semi-fluid portion of the cytoplasm); others are bound to the rough endoplasmic reticulum, giving it the appearance of roughness and thus its name, or to the nuclear envelope. As ribozymes are partly constituted from RNA, it is thought that they might be remnants of the RNA world. Although catalysis of the peptide bond involves the C2 hydroxyl of RNA's P-site adenosine in a protein shuttle mechanism, other steps in protein synthesis (such as translocation) are caused by changes in protein conformations.

Ribosomes are sometimes referred to as organelles, but the use of the term *organelle* is often restricted to describing sub-cellular components that include a phospholipid membrane, which ribosomes, being entirely particulate, do not. For this reason, ribosomes may sometimes be described as "non-membranous organelles".

Ribosomes were first observed in the mid-1950s by Romanian cell biologist George Palade using an electron microscope as dense particles or granules for which he would win the Nobel Prize. The term "ribosome" was proposed by scientist Richard B. Roberts in 1958:

> *During the course of the symposium a semantic difficulty became apparent. To some of the participants, "microsomes" mean the ribonucleoprotein particles of the microsome fraction contaminated by other protein and lipid material; to others, the microsomes consist of protein and lipid contaminated by particles. The phrase "microsomal particles" does not seem adequate, and "ribonucleoprotein particles of the microsome fraction" is much too awkward. During the meeting, the word "ribosome" was suggested, which has a very satisfactory name and a pleasant sound. The present confusion would be eliminated if "ribosome" were adopted to designate ribonucleoprotein particles in sizes ranging from 35 to 100S.*
>
> *– Roberts, R. B., Microsomal Particles and Protein Synthesis*

The structure and function of the ribosomes and associated molecules, known as the *translational apparatus*, has been of research interest since the mid-twentieth century and is a very active field of study today.

Ribosomes consist of two subunits that fit together and work as one to translate the mRNA into a polypeptide chain during protein synthesis. Bacterial subunits consist of one or two and eukaryotic of one or three very large RNA molecules (known as ribosomal RNA or rRNA) and multiple smaller protein molecules. Crystallographic work has shown that there are no ribosomal proteins close to the reaction site for polypeptide synthesis. This suggests that the protein components of ribosomes act as a scaffold that may enhance the ability of rRNA to synthesize protein rather than directly participating in catalysis.

Biogenesis

In bacterial cells, ribosomes are synthesized in the cytoplasm through the transcription of multiple ribosome gene operons. In eukaryotes, the process takes place both in the cell cytoplasm and in the nucleolus, which is a region within the cell nucleus. The assembly process involves the coordinated function of over 200 proteins in the synthesis and processing of the four rRNAs, as well as assembly of those rRNAs with the ribosomal proteins.

Ribosome Locations

Ribosomes are classified as being either "free" or "membrane-bound".

Free and membrane-bound ribosomes differ only in their spatial distribution; they are identical in structure. Whether the ribosome exists in a free or membrane-bound state depends on the presence of an ER-targeting signal sequence on the protein being synthesized, so an individual ribosome might be membrane-bound when it is making one protein, but free in the cytosol when it makes another protein.

Free Ribosomes

Free ribosomes can move about anywhere in the cytosol, but are excluded from the cell nucleus and other organelles. Proteins that are formed from free ribosomes are released into the cytosol and used within the cell. Since the cytosol contains high concentrations of glutathione and is, therefore, a reducing environment, proteins containing disulfide bonds, which are formed from oxidized cysteine residues, cannot be produced in this compartment.

Membrane-bound Ribosomes

When a ribosome begins to synthesize proteins that are needed in some organelles, the ribosome making this protein can become "membrane-bound". In eukaryotic cells this happens in a region of the endoplasmic reticulum (ER) called the "rough ER". The newly produced polypeptide chains are inserted directly into the ER by the ribosome and are then transported to their destinations, through the secretory pathway. Bound ribosomes usually produce proteins that are used within the plasma membrane or are expelled from the cell via *exocytosis.*

Structure

The ribosomal subunits of prokaryotes and eukaryotes are quite similar. The unit of measurement is the Svedberg unit, a measure of the rate of sedimentation in centrifugation rather than size and accounts for why fragment names do not add up (70S is made of 50S and 30S). Prokaryotes have 70S ribosomes, each consisting of a small (30S) and a large (50S) subunit. Their large subunit is composed of a 5S RNA subunit (consisting of 120 nucleotides), a 23S RNA subunit (2900 nucleotides) and 34 proteins. The 30S subunit has a 1540 nucleotide RNA subunit (16S) bound to 21 proteins.

Eukaryotes have 80S ribosomes, each consisting of a small (40S) and large (60S) subunit. Their large subunit is composed of a 5S RNA (120 nucleotides), a 28S RNA (4700 nucleotides), a 5.8S subunit (160 nucleotides) and ~49 proteins. The 40S subunit has a 1900 nucleotide (18S) RNA and ~33 proteins.

The ribosomes found in chloroplasts and mitochondria of eukaryotes also consist of large and small subunits bound together with proteins into one 70S particle. These organelles are believed to be descendants of bacteria and as such their ribosomes are similar to those of bacteria.

The various ribosomes share a core structure, which is quite similar despite the large differences in size. Much of the RNA is highly organized into various tertiary structural motifs, for example pseudoknots that exhibit coaxial stacking. The extra RNA in the larger ribosomes is in several long continuous insertions, such that they form loops out of the core structure without disrupting or changing it. All of the catalytic activity of the ribosome is carried out by the RNA; the proteins reside on the surface and seem to stabilize the structure.

The differences between the bacterial and eukaryotic ribosomes are exploited by pharmaceutical chemists to create antibiotics that can destroy a bacterial infection without harming the cells of the infected person. Due to the differences in their structures, the bacterial 70S ribosomes are vulnerable to these antibiotics while the eukaryotic 80S ribosomes are not. Even though mitochondria possess ribosomes similar to the bacterial ones, mitochondria are not affected by these antibiotics because they are surrounded by a double membrane that does not easily admit these antibiotics into the organelle.

High-resolution Structure

The general molecular structure of the ribosome has been known since the early 1970s. In the early 2000s the structure has been achieved at high resolutions, on the order of a few ångströms. The first papers giving the structure of the ribosome at atomic resolution were published in rapid succession in late 2000. First, the 50S (large prokaryotic) subunit from the archaeon *Haloarcula marismortui* was published. Soon after, the structure of the 30S subunit from *Thermus thermophilus* was published. Shortly thereafter, a more detailed structure was published. These structural studies were awarded the Nobel Prize in Chemistry in 2009. Early the next year (May 2001) these coordinates were used to reconstruct the entire *T. thermophilus* 70S particle at 5.5 ångström resolution.

Two papers were published in November 2005 with structures of the *Escherichia coli* 70S ribosome. The structures of a vacant ribosome were determined at 3.5-angström resolution using x-ray crystallography. Then, two weeks later, a structure based on cryo-electron microscopy was published, which depicts the ribosome at 11-15 angström resolution in the act of passing a newly synthesized protein strand into the protein-conducting channel.

First atomic structures of the ribosome complexed with tRNA and mRNA molecules were solved by using X-ray crystallography by two groups independently, at 2.8 angstrom and at 3.7 angstrom. These structures allow one to see the details of interactions of the *Thermus thermophilus* ribosome with mRNA and with tRNAs bound at classical ribosomal sites. Interactions of the ribosome with long mRNAs containing Shine-Dalgarno sequences were visualized soon after that at 4.5- to 5.5-angström resolution.

Function

Ribosomes are the workhorses of protein biosynthesis, the process of translating mRNA into protein. The mRNA comprises a series of codons that dictate to the ribosome the sequence of the amino acids needed to make the protein. Using the mRNA as a template, the ribosome traverses each codon (3 nucleotides) of the mRNA, pairing it with the appropriate amino acid provided by a tRNA. Molecules of transfer RNA (tRNA) contain a complementary anticodon on one end and the appropriate amino acid on the other. The small ribosomal subunit, typically bound to a tRNA containing the amino acid methionine, binds to an AUG codon on the mRNA and recruits the large ribosomal subunit. The ribosome then contains three RNA binding sites, designated A, P and E. The A site binds an aminoacyl-tRNA (a tRNA bound to an amino acid); the P site binds a peptidyl-tRNA (a tRNA bound to the peptide being synthesized); and the E site binds a free tRNA before it exits the ribosome. Protein synthesis begins at a start codon AUG near the 5' end of the mRNA. mRNA binds to the P site of the ribosome first. The ribosome is able to identify the start codon by use of the Shine-Dalgarno sequence of the mRNA in prokaryotes and Kozak box in eukaryotes.

The ribosome uses tRNA that matches the current codon (triplet) on the mRNA to append an amino acid to the polypeptide chain. This is done for each triplet on the mRNA, while the ribosome moves towards the 3' end of the mRNA. Usually in bacterial cells, several ribosomes are working parallel on a single mRNA, forming what is called a *polyribosome* or *polysome*.

Cytoskeleton: Microtubules, Microfilaments and Microvilli

The cytoskeleton (also CSK) is a cellular "scaffolding" or "skeleton" contained within the cytoplasm and is made out of protein. The cytoskeleton is present in all cells; it was once thought to be unique to eukaryotes, but recent research has identified the prokaryotic cytoskeleton. It is a dynamic structure that maintains cell shape, protects the cell, enables cellular motion (using structures such as flagella, cilia and lamellipodia), and plays important roles in both intracellular transport (the movement of vesicles and organelles, for example) and cellular division. The concept of a protein mosaic that dynamically coordinated cytoplasmic biochemistry was proposed by Rudolph Peters in 1929 while the term (*cytosquelette*, in French) was first introduced by French embryologist Paul Wintrebert in 1931.

The Eukaryotic Cytoskeleton

Eukaryotic cells contain three main kinds of cytoskeletal filaments, which are microfilaments, intermediate filaments, and microtubules. The cytoskeleton provides the cell with structure and shape, and by excluding macromolecules from some of the cytosol it adds to the level of macromolecular crowding in this compartment. Cytoskeletal elements interact extensively and intimately with cellular membranes.

Microfilament

Microfilaments (or actin filaments) are the thinnest filaments of the cytoskeleton found in the cytoplasm of all eukaryotic cells. These linear polymers of actin subunits are flexible and relatively strong, resisting buckling by multi-piconewton compressive forces and filament fracture by nanonewton tensile forces. Microfilaments are highly versatile, functioning in (a) cell crawling, amoeboid movement, and changes in cell shape, where one end of the actin filament elongates while the other end contractile, presumably by myosin II molecular motors (though an alternative "actoclampin"-driven expansile molecular motors exists), and (b) actomyosin-driven contractile molecular motors, where the *thin filaments* serve as tensile platforms for myosin's ATP hydrolysis-dependent pulling action in muscle contraction and uropod advancement.

Organization

Actin filaments are assembled in two general types of structures: bundles and networks. Bundles can be composed of polar (all barbed ends point to the same end of the bundle) or non-polar (barbed ends point towards both ends) arrays of filaments. A class of Actin-binding proteins, called cross-linking proteins, dictate the formation of these structures. Cross-linking proteins determine filament orientation and spacing in the bundles and networks. These structures are regulated by many other classes of actin-binding proteins (motor proteins, branching proteins, severing proteins, polymerization promoters, capping proteins, etc.).

In Vitro *Self-assembly*

The thinnest fibres of the cytoskeleton (measuring approximately 6 nm in diameter), microfilaments are formed by the head-to-tail polymerization of actin monomers (also known as globular or G-actin). Actin subunits as part of a fibre are referred to as filamentous actin (or F-actin). Each microfilament is made up of two helical interlaced strands of subunits. Much like microtubules, actin filaments are polarized, with their fast-growing barbed-ends (because of their appearance in electron micrographs after binding of myosin S1 sub-fragments) and a slow-growing pointed-end (again based on the pattern created by S1 binding). The pointed end is sometimes referred to as the minus (-) end and the barbed end is sometimes referred to as the plus (+) end because of the growth rates, but this is nomenclature adapted from the microtubule field, and is not generally accepted in the actin field.

In vitro actin polymerization, nucleation, starts with the self-association of three G-actin monomers to form a trimer. ATP-actin then binds the barbed end, and the ATP is subsequently hydrolyzed with a half time of about 2 seconds and the inorganic phosphate released with a half-time of about 6 minutes, which reduces the binding strength between neighbouring units and generally destabilizes the filament. *In vivo* actin polymerization is catalyzed by a new class of filament end-tracking molecular motors known as actoclampins. Recent evidence suggests that ATP hydrolysis can be prompt in such cases (i.e., the rate of monomer incorporation is matched by the rate of ATP hydrolysis).

ADP-actin dissociates slowly from the pointed end, but this process is greatly accelerated by ADP-cofilin, which severs ADP-rich regions nearest the (–)-ends. Upon release, ADP-actin undergoes exchange of its bound ADP for solution-phase ATP, thereby forming the ATP-actin monomeric units needed for further barbed-end filament elongation. This rapid turnover is important for the cell's movement. End-capping proteins such as CapZ prevent the addition or loss of monomers at the filament end where actin turnover is unfavourable like in the muscle apparatus.

Mechanism of Force Generation

As a result of the ATP hydrolysis, filaments elongate approximately 10 times faster at their barbed ends than their pointed ends. At steady-state, the polymerization rate at the barbed end matches the depolymerization rate at the pointed end, and microfilaments are said to be treadmilling. Treadmilling results in elongation in the barbed end and shortening in the pointed-end, so that the filament in total moves. Since both processes are energetically favourable, this means force is generated, and the energy ultimately comes from the NTP hydrolization.

Actin in Cells

Intracellular actin cytoskeletal assembly and disassembly are tightly regulated by cell signalling mechanisms. Many signal transduction systems use the actin cytoskeleton as a scaffold holding them at or near the inner face of the peripheral membrane. This subcellular location allows immediate and exquisite responsiveness to transmembrane receptor action and signal-processing enzyme cascades. Because actin monomers must be recycled to sustain high rates of actin-based motility during chemotaxis, cell signalling is believed to activate cofilin, an actin-filament depolymerizing protein which binds to ADP-rich actin subunits nearest the filament's pointed-end and promotes filament fragmentation, with concomitant depolymerization to liberate actin monomers. The protein profilin enhances the ability of monomers to assemble by stimulating the exchange of actin-bound ADP for solution-phase ATP to yield Actin-ATP and ADP. In most animal cells, monomeric actin is bound to profilin and thymosin-beta4, both of which preferentially bind with one-to-one stoichiometry to ATP-containing monomers. Although thymosin-beta4 is strictly a monomer-sequestering protein, the behavior of profilin is far more complex. Profilin is transferred to the leading edge by virtue of its PIP2 binding site, and profilin also employs its poly-L-proline binding site to dock onto end-tracking proteins. Once bound, Profilin-Actin-ATP is loaded into the monomer-insertion site of actoclampin motors. Another important component in filament formation is the Arp2/3 complex, which binds to the side of an already existing filament (or "mother filament"), where it nucleates the formation of a new actin filament and creates a fan-like branched filament network.

In non-muscle cells, actin filaments are formed at/near membrane surfaces. Their formation and turnover are regulated by many proteins, including

- Filament end-tracking protein (e.g., formins, VASP, N-WASP)
- Filament-nucleator known as the Actin-Related Protein-2/3 (or Arp2/3) complex
- Filament cross-linkers (e.g., α-actinin and fascin)
- Actin monomer-binding proteins profilin and thymosin-α4

- Filament barbed-end cappers such as Capping Protein and CapG, *etc.*
- Filament-severing proteins like gelsolin
- pointed-End depolymerizing proteins such as ADF/cofilin.

The actin filament network in non-muscle cells is highly dynamic. As first proposed by Dickinson & Purich (*Biophysical Journal* 92: 622-631), the actin filament network is arranged with the barbed-end of each filament attached to the cell's peripheral membrane by means of clamped-filament elongation motors ("actoclampins") formed from a filament barbed-end and a clamping protein (formins, VASP, Mena, WASP, and N-WASP). The primary substrate for these elongation motors is Profilin-Actin-ATP complex which is directly transferred to elongating filament ends (Dickinson, Southwick & Purich, 2002). The pointed-end of each filament is oriented toward the cell's interior. In the case of lamellipodial growth, the Arp2/3 complex generates a branched network, and in filopodia a parallel array of filaments is formed.

Actin Acts as a Track for Myosin Motor Motility

Myosin motors are intracellular ATP-dependent enzymes that bind to and move along actin filaments. Various classes of myosin motors have very different behaviors, including exerting tension in the cell and transporting cargo vesicles.

A Proposed Model - Actoclampins Track Filament Ends

One proposed model suggests the existence of actin filament barbed-end-tracking molecular motors termed actoclampin. The proposed actoclampins generate the propulsive forces needed for actin-based motility of lamellipodia, filopodia, invadipodia, dendritic spines, intracellular vesicles, and motile processes in endocytosis, exocytosis, podosome formation, and phagocytosis. Actoclampin motors also propel such intracellular pathogens as *Listeria monocytogenes*, *Shigella flexneri*, *Vaccinia* and *Rickettsia*. When assembled under suitable conditions, these end-tracking molecular motors can also propel biomimetic particles.

The term actoclampin is derived from *acto-* to indicate the involvement of an actin filament, as in actomyosin, and *clamp* to indicate a clasping device used for strengthening flexible/moving objects and for securely fastening two or more components, followed by the suffix *-in* to indicate its protein origin. An actin filament end-tracking protein may thus be termed a clampin.

Dickinson and Purich (2002) recognized that prompt ATP hydrolysis could explain the forces achieved during actin-based motility. They proposed a simple mechanoenzymatic sequence known as the Lock, Load & Fire Model, in which an end-tracking protein remains tightly bound ("locked" or clamped) onto the end of one sub-filament of the double-stranded actin filament. After binding to Glycyl-Prolyl-Prolyl-Prolyl-Prolyl-Prolyl-registers on tracker proteins, Profilin-ATP-actin is delivered ("loaded") to the unclamped end of the other sub-filament, whereupon ATP within the already clamped terminal subunit of the other subfragment is hydrolyzed ("fired"), providing the energy needed to release that arm of the end-tracker, which then can bind another Profilin-ATP-actin to begin a new monomer-addition round.

The following steps describe one force-generating cycle of an actoclampin molecular motor:

1. The polymerization cofactor profilin and the ATP -actin combine to form a profilin-ATP-actin complex that then binds to the end-tracking unit
2. The cofactor and monomer are transferred to the barbed-end of an actin already clamped filament
3. The tracking unit and cofactor dissociate from the adjacent protofilament, in a step that can be facilitated by ATP hydrolysis energy to modulate the affinity of the cofactor and/or the tracking unit for the filament; and this mechanoenzymatic cycle is then repeated, starting this time on the other sub-filament growth site.

When operating with the benefit of ATP hydrolysis, AC motors generate per-filament forces of 8–9 pN, which is far greater than the per-filament limit of 1–2 pN for motors operating without ATP hydrolysis (Dickinson and Purich, 2002, 2006; Dickinson, Caro and Purich, 2004). The term actoclampin is generic and applies to all actin filament end-tracking molecular motors, irrespective of whether they are driven actively by an ATP-activated mechanism or passively.

Some actoclampins (e.g., those involving Ena/VASP proteins, WASP, and N-WASP) apparently require Arp2/3-mediated filament initiation to form the actin polymerization nucleus that is then "loaded" onto the end-tracker before processive motility can commence. To generate a new filament, Arp2/3 requires a "mother" filament, monomeric ATP-actin, and an activating domain from Listeria ActA or the VCA region of N-WASP. Ther Arp2/3 complex binds to the side of the mother filament, forming a Y-shaped branch having a 70 degree angle with respect to the longitudinal axis of the mother filament. Then upon activation by ActA or VCA, the Arp complex is believed to undergo a major conformational change, bringing its two actin-related protein subunits near enough to each other to generate a new filament gat. Whether ATP hydrolysis may be required for nucleation and/or Y-branch release is a matter under active investigation.

Actin

Actin is a globular, roughly 42-kDa protein found in all eukaryotic cells (the only known exception being nematode sperm) where it may be present at concentrations of over 100 μM. It is also one of the most highly-conserved proteins, differing by no more than 20% in species as diverse as algae and humans. Actin is the monomeric subunit of two types of filaments in cells: microfilaments, one of the three major components of the cytoskeleton, and thin filaments, part of the contractile apparatus in muscle cells.

Thus, actin participates in many important cellular processes including muscle contraction, cell motility, cell division and cytokinesis, vesicle and organelle movement, cell signalling, and the establishment and maintenance of cell junctions and cell shape. Many of these processes are mediated by extensive and intimate interactions of actin with cellular membranes. In vertebrates, three main groups of actin isoforms, alpha, beta, and gamma have been identified. The alpha actins are found in muscle tissues and are a major constituent of the contractile apparatus. The beta and gamma actins co-exist in most cell types as components of the cytoskeleton, and as mediators of internal cell motility.

Formation of Thin Filament

Principal interactions of structural proteins are at cadherin-based adherens junction. Actin filaments are linked to α-actinin and to the membrane through vinculin. The head domain of vinculin associates to E-cadherin via á-, â-, and β-catenins. The tail domain of vinculin binds to membrane lipids and to actin filaments.

The protein actin is one of the most highly conserved throughout evolution because it interacts with a large number of other proteins, with 80.2% sequence conservation at the gene level between *Homo sapiens* and *Saccharomyces cerevisiae* (a species of yeast), and 95% conservation of the primary structure of the protein product.

Although most yeasts have only a single actin gene, higher eukaryotes, in general, express several isoforms of actin encoded by a family of related genes. Mammals have at least six actin isoforms coded by separate genes, which are divided into three classes (alpha, beta and gamma) according to their isoelectric point. In general, alpha actins are found in muscle, whereas beta and gamma isoforms are prominent in non-muscle cells (α- and β1-cytoplasmic). Although the amino acid sequences and *in vitro* properties of the isoforms are highly similar, these isoforms cannot completely substitute for one another *in vivo*.

The typical actin gene has an approximately 100-nucleotide 5' UTR, a 1200-nucleotide translated region, and a 200-nucleotide 3' UTR. The majority of actin genes are interrupted by introns, with up to 6 introns in any of 19 well-characterised locations. The high conservation of the family makes actin the favoured model for studies comparing the introns-early and introns-late models of intron evolution.

All non-spherical prokaryotes appear to possess genes such as MreB, which encode homologues of actin; these genes are required for the cell's shape to be maintained. The plasmid-derived gene ParM encodes an actin-like protein whose polymerised form is dynamically unstable, and appears to partition the plasmid DNA into the daughter cells during cell division by a mechanism analogous to that employed by microtubules in eukaryotic mitosis. Actin is found in both smooth and rough endoplasmic reticulums.

Functions

Actin has four main functions in cells:

- To form microfilaments, the most dynamic of the three subclasses of the cytoskeleton, which give mechanical support to cells, and hardwire the cytoplasm with the surroundings to support signal transduction.
- To allow cell motility, including phagocytosis of bacteria by macrophages.
- In muscle cells, to be the scaffold on which myosin proteins generate force to support muscle contraction.
- In non-muscle cells, to be a track for cargo transport myosins (non-conventional myosins) such as myosin V and VI. Non-conventional myosins use ATP hydrolysis to transport cargo, such as vesicles and organelles, in a directed fashion much faster than diffusion. Myosin V walks towards the barbed end of actin filaments, while myosin VI walks toward the pointed end. Most actin filaments are arranged with the barbed end toward the cellular membrane and the pointed end toward the cellular interior. This arrangement allows myosin V to be an effective

motor for export of cargos, and myosin VI to be an effective motor for import.

Directionality

The polarity of an actin filament can be determined by decorating the microfilament with myosin "S1" fragments, creating barbed (+) and pointed (-) ends on the filament. An S1 fragment is composed of the head and neck domains of myosin II. Under physiologic conditions, G-actin (the monomer form) is transformed to F-actin (the polymer form) by ATP, where the role of ATP is essential.

Nucleation and Polymerization

Actin polymerization and depolymerization is necessary in chemotaxis and cytokinesis. Nucleating factors are necessary to stimulate actin polymerization. One such nucleating factor is the ARP complex which acts as a barbed end of actin in its shape to stimulate the nucleation of G-actin (or monomeric actin). The Arp2/3 complex can also bind to actin filaments at 70 degrees to form new actin branches off of existing actin filaments. Also, actin filaments themselves bind ATP, and hydrolysis of this ATP stimulates destabilization of the polymer. The growth of actin filaments can be regulated by thymosin and profilin. Thymosin binds to G-actin to prevent it from polymerizing while profilin binds to G-actin to promote monomeric addition to the barbed, plus end.

Microfilaments

Individual subunits of microfilaments are known as globular actin (G-actin). G-actin subunits assemble into long filamentous polymers called F-actin. Two parallel F-actin strands must rotate 166 degrees to layer correctly on top of each other. This creates the double helix structure of the microfilaments of the cytoskeleton. Microfilaments measure approximately 7 nm in diameter with a loop of the helix repeating every 37 nm.

Actomyosin Filaments

In muscle, actin is the major component of *thin filaments*, which, together with the motor protein myosin (which forms *thick filaments*), are arranged into actomyosin myofibrils. These fibrils comprise the mechanism of muscle contraction. Using the hydrolysis of ATP for energy, myosin heads undergo a cycle during which they attach to thin filaments, exerting a tension, and then depending on the load, perform a power stroke that causes the thin filaments to slide past, shortening the muscle.

In contractile bundles, the actin-bundling protein alpha-actinin separates each thin filament by ~35 nm. This increase in distance allows thick filaments to fit in between and interact, enabling deformation or contraction. In deformation, one end of myosin is bound to the plasma membrane while the other end "walks" toward the plus end of the actin filament. This pulls the membrane into a different shape relative to the cell cortex.

For contraction, the myosin molecule is usually bound to two separate filaments and both ends simultaneously "walk" toward their filament's plus end, sliding the actin filaments closer to each other. This results in the shortening, or contraction, of the actin bundle (but not the filament). This mechanism is responsible for muscle contraction and cytokinesis, the division of one cell into two.

Nuclear Actin

Actin is essential for transcription from RNA polymerases I, II and III. In Pol I transcription, actin and myosin (MYO1C, which binds DNA) act as a molecular motor. For Pol II transcription, â-actin is needed for the formation of the pre-initiation complex. Pol III contains α-actin as a subunit. Actin can also be a component of chromatin remodeling complexes as well as pre-mRNP particles (that is, precursor messenger RNA bundled in proteins), and is involved in nuclear export of RNAs and proteins.

History

Actin was first observed experimentally in 1887 by W.D. Halliburton, who extracted a protein from muscle that 'coagulated' preparations of myosin, which he dubbed "myosin-ferment." However, Halliburton was unable to further characterise his findings, and the discovery of actin is credited instead to Brúnó F. Straub, a young biochemist working in Albert Szent-Györgyi's laboratory at the Institute of Medical Chemistry at the University of Szeged, Hungary.

In 1942, Straub developed a novel technique for extracting muscle protein that allowed him to isolate substantial amounts of relatively pure actin. Straub's method is essentially the same as that used in laboratories today.

Szent-Gyorgyi had previously described the more viscous form of myosin produced by slow muscle extractions as 'activated' myosin, and, since Straub's protein produced the activating effect, it was dubbed *actin*. The hostilities of World War II meant that Szent-Gyorgyi and Straub were unable to publish the work in Western scientific journals; it became well-known in the West only in 1945, when it was published as a supplement to the Acta Physiologica Scandinavica. Straub continued to work on actin, and in 1950 reported that actin contains bound ATP and that, during polymerisation of the protein into microfilaments, the nucleotide is hydrolysed to ADP and inorganic phosphate (which remain bound in the microfilament). Straub suggested that the transformation of ATP-bound actin to ADP-bound actin played a role in muscular contraction. In fact, this is true only in smooth muscle, and was not supported through experimentation until 2001.

The crystal structure of G-actin was solved in 1990 by Kabsch and colleagues. In the same year, a model for F-actin was proposed by Holmes and colleagues. The model was derived by fitting a helix of G-actin structures according to low-resolution fibre diffraction data from the filament. Several models of the filament have been proposed since. However, there is still no high-resolution X-ray structure of F-actin. The Listeria bacteria use the cellular machinery to move around inside the host cell, by inducing directed polymerisation of actin by the ActA transmembrane protein, thus pushing the bacterial cell around.

Intermediate Filaments

Intermediate filaments (IFs) are a family of related proteins that share common structural and sequence features. Intermediate filaments have an average diameter of 10 nanometers, which is between that of actin (microfilaments) and microtubules, although they were initially designated 'intermediate' because their average diameter was between those of narrower microfilaments (actin) and wider myosin filaments. Most types of intermediate filaments are cytoplasmic, but one type, the lamins, are nuclear.

Structure

The domain structure of IF molecules is conserved. Each protein has a non-alpha-helical (globular) domain at the N and C-termini which surrounds the alpha-helical rod domain. The basic building block for IFs is a parallel and in register dimer. The dimer is formed through the interaction of the rod domain to form a coiled coil. Cytoplasmic IF assemble into non-polar unit-length filaments (ULF) which then assemble into longer structures. Part of the assembly process includes a compaction step, in which ULF tighten and assume a smaller diameter. The reasons for this compaction are not well understood, and IF are routinely observed to have diameters ranging between 6 and 12nm. The anti-parallel orientation of tetramers means that, unlike microtubules and microfilaments which have a plus end and a minus end, IFs lack polarity.

Also, as opposed to actin or tubulin, intermediate filaments do not contain a binding site for a nucleoside triphosphate. Cytoplasmic IF do not undergo treadmilling like microtubules and actin fibres, but they are dynamic.

Biomechanical Properties

IFs are rather deformable proteins that can be stretched several times their initial length. The key to facilitate this large deformation is due to their hierarchical structure, which facilitates a cascaded activation of deformation mechanisms at different levels of strain.

Types

There are about 70 different genes coding for various intermediate filament proteins. However, different kinds of IFs share basic characteristics: they are all polymers that generally measure between 9-11 nm in diameter when fully assembled. IF are subcategorized into six types based on similarities in amino acid sequence and protein structure.

Types I and II - Acidic and Basic Keratins

These proteins are the most diverse among IFs and constitute type I (acidic) and type II (basic) IF proteins. The many isoforms are divided in two groups:

- epithelial keratins (about 20) in epithelial cells (image to right)
- trichocytic keratins (about 13) (hair keratins) which make up hair, nails, horns and reptilian scales.

Regardless of the group, keratins are either acidic or basic. Acidic and basic keratins bind each other to form acidic-basic heterodimers and these heterodimers then associate to make a keratin filament.

Type III

There are four proteins classed as type III IF proteins which may form homo- or heteropolymeric proteins.

- Desmin IFs are structural components of the sarcomeres in muscle cells.
- GFAP (glial fibrillary acidic protein) is found in astrocytes and other glia.
- Peripherin found in peripheral neurons.
- Vimentin, the most widely distributed of all IF proteins, can be found in fibroblasts, leukocytes, and blood vessel endothelial cells. They support the cellular membranes and keep some organelles in a fixed place within the cytoplasm.

Type IV

- α-Internexin
- Neurofilaments - the type IV family of intermediate filaments that is found in high concentrations along the axons of vertebrate neurons.
- Synemin
- Syncoilin.

Type V - Nuclear Lamins

Lamins : Lamins are fibrous proteins having structural function in the cell nucleus.

In metazoan cells there are A and B type lamins which differ in their length and pI. Human cells have three differentially regulated genes. B-type lamins are present in every cell. B type lamins, B1 and B2, are expressed from the LMNB1 and LMNB2 genes on 5q23 and 19q13, respectively. A-type lamins are only expressed following gastrulation. Lamin A and C are the most common A-type lamins and are splice variants of the LMNA gene found at 1q21.

These proteins localize to two regions of the nuclear compartment, the nuclear lamina — a proteinaceous structure layer subjacent to the inner surface of the nuclear envelope and throughout the nucleoplasm in the nucleoplasmic "veil".

Comparison of the lamins to vertebrate cytoskeletal IFs shows that lamins have an extra 42 residues (six heptads) within coil 1b. The c-terminal tail domain contains a nuclear localization signal (NLS), an Ig-fold like domain, and in most cases a carboxy-terminal CaaX box that is isoprenylated and carboxymethylated (lamin C does not have a CAAX box). Lamin A is further processed to remove the last 15 amino acids and its farnesylated cysteine.

During mitosis, lamins are phosphorylated by MPF which drives the disassembly of the lamina and the nuclear envelope.

Cell Adhesion

At the plasma membrane, some keratins interact with desmosomes (cell-cell adhesion) and hemidesmosomes (cell-matrix adhesion) via adapter proteins.

Associated Proteins

Filaggrin binds to keratin fibres in epidermal cells. Plectin links vimentin to other vimentin fibres, as well as to microfilaments, microtubules, and myosin II.

Keratin filaments in epithelial cells link to desmosomes (desmosomes connect the cytoskeleton together) through plakoglobin, desmoplakin, desmogleins and desmocollins; desmin filaments are connected in a similar way in heart muscle cells.

Diseases Arising from Mutations in IF Genes

- Epidermolysis bullosa simplex; K5 or K14 mutation
- Laminopathies are a family of diseases caused by mutations in nuclear lamins and include Hutchinson Gilford Progeria Syndrome and various lipodystrophies and cardiomyopathies among others.
- Human Intermediate Filament Database(HIFD), a comprehensive database of human intermediate filament proteins, their associated variations and diseases.
- envelope.

Microtubules

Microtubules are one of the components of the cytoskeleton. They have a diameter of 25 nm and length varying from 200 nanometers to 25 micrometers. Microtubules serve as structural components within cells and are involved in many cellular processes including mitosis, cytokinesis, and vesicular transport.

Structure

Microtubules are polymers of α- and β-tubulin dimers. The tubulin dimers polymerize end to end in protofilaments. The protofilaments then bundle into hollow cylindrical filaments. Typically the protofilaments arrange themselves in an imperfect helix with one turn of the helix containing 13 tubulin dimers each from a different protofilament. The image above illustrates a small section of microtubule, a few αβ dimers in length.

Another important feature of microtubule structure is polarity. Tubulin polymerizes end to end with the α subunit of one tubulin dimer contacting the β subunit of the next. Therefore, in a protofilament, one end will have the α subunit exposed while the other end will have the α subunit exposed. These ends are designated the (“) and (+) ends, respectively. The protofilaments bundle parallel to one another, so in a microtubule, there is one end, the (+) end, with only α subunits exposed while the other end, the (“) end, only has β subunits exposed. The (-) end is capped so elongation of the microtubule occurs from the (+) direction.

Organization within Cells

Microtubules are nucleated and organized by the microtubule organizing centres (MTOCs), such as centrioles and basal bodies. Microtubules are part of a structural network (the cytoskeleton) within the cell's cytoplasm, but, in addition to structural support, microtubules take part in many other processes, as well. They are capable of growing and shrinking in order to generate force, and there are also motor proteins that allow organelles and other cellular factors to move along the microtubule. A notable structure involving microtubules is the mitotic spindle used by eukaryotic cells to segregate their chromosomes correctly during cell division. Microtubules are also part of the cilia and flagella of eukaryotic cells (prokaryote flagella are entirely different).

The process of mitosis is facilitated by a subgroup of microtubules known as astral microtubules, defined as a microtubule originating from the centrosome that does not connect to a kinetochore. Astral microtubules develop in the actin skeleton and interact with the cell cortex to aid in spindle orientation. They are organized into radial arrays around the centrosomes. The turn-over rate of this population of microtubules is higher than any other population. Astral microtubules function in concert with specialized dynein motors, which are oriented with the light chain portion attached to the cell membrane and the dynamic portion attached to the microtubule. This allows for dynein contraction to pull the centrosome towards the cell membrane, thus assisting in cytokinesis. Astral microtubules are not required for the progression of mitosis, but they are required to ensure the fidelity of the process; they are required for the correct positioning and orientation of the mitotic spindle apparatus. They are also involved in determination of cell division site based on the geometry and polarity of the cells. The maintenance of astral microtubules is dependent on the integrity of centrosome. It is also dependent on several microtubule-associated proteins such as EB1 and Adenomatous Polyposis Coli (APC).

Nucleation and Growth

Polymerization of microtubules is nucleated in a microtubule organizing centre. Contained within the MTOC is another type of tubulin, β-tubulin, which is distinct from the α and β subunits which compose the microtubules themselves. The α-tubulin combines with several other associated proteins to form a circular structure known as the "α-tubulin ring complex." This complex acts as a scaffold for α/β tubulin dimers to begin polymerization; it acts as a cap of the (") end while microtubule growth continues away from the MTOC in the (+) direction.

Dynamic Instability

During polymerization, both the α- and β-subunits of the tubulin dimer are bound to a molecule of GTP. While the GTP bound to α-tubulin is stable, the GTP bound to α-tubulin may be hydrolyzed to GDP shortly after assembly. The kinetics of GDP-tubulin are different from those of GTP-tubulin; GDP-tubulin is prone to depolymerization.

A GDP-bound tubulin subunit at the tip of a microtubule will fall off, though a GDP-bound tubulin in the middle of a microtubule cannot spontaneously pop out. Since tubulin adds onto the end of the microtubule only in the GTP-bound state, there is generally a cap of GTP-bound tubulin at the tip of the microtubule, protecting it from disassembly. When hydrolysis catches up to the tip of the microtubule, it begins a rapid depolymerization and shrinkage. This switch from growth to shrinking is called a catastrophe. GTP-bound tubulin can begin adding to the tip of the microtubule again, providing a new cap and protecting the microtubule from shrinking. This is referred to as rescue.

Chemical Effects on Microtubule Dynamics

Microtubule dynamics can also be altered by drugs.

- For example, the taxane drug class (e.g. paclitaxel or docetaxel), used in the treatment of cancer, blocks dynamic instability by stabilizing GDP-bound tubulin in the microtubule. Thus, even when hydrolysis of GTP reaches the tip of the microtubule, there is no depolymerization and the microtubule does not shrink back.
- Nocodazole and Colchicine have the opposite effect, blocking the polymerization of tubulin into microtubules.

Motor Proteins

In addition to movement generated by the dynamic instability of the microtubule itself, the fibres are substrates along which motor proteins can move. The major microtubule motor proteins are kinesin, which generally moves towards the (+) end of the microtubule, and dynein, which generally moves towards the (“) end.

Postulated Role in Consciousness

In their controversial Orch-OR theory of consciousness, Roger Penrose and Stuart Hameroff postulate that microtubules in neurons conduct quantum-level manipulations of matter which produces consciousness, based partially on widely recorded observations of Gamma Synchrony that indicate that information may propagate through the brain much faster than a chemically mediated neural network would physically permit. Max Tegmark disputes the relevance of these observations, and the matter remains open to debate.

The Prokaryotic Cytoskeleton

The prokaryotic cytoskeleton is the collective name for all structural filaments in prokaryotes. It was once thought that prokaryotic cells did not possess cytoskeletons, but recent advances in visualization technology and structure determination have shown that filaments indeed exist in these cells. In fact, analogues for all major cytoskeletal proteins in eukaryotes have been found in prokaryotes. Cytoskeletal elements play essential roles in cell division, protection, shape determination, and polarity determination in various prokaryotes.

The cytoskeleton was previously thought to be a feature only of eukaryotic cells, but homologues to all the major proteins of the eukaryotic cytoskeleton have recently been found in prokaryotes. Although the evolutionary relationships are so distant that they are not obvious from protein sequence comparisons alone, the similarity of their three-dimensional structures and similar functions in maintaining cell shape and polarity provides strong evidence that the eukaryotic and prokaryotic cytoskeletons are truly homologous. However, some structures in the bacterial cytoskeleton may have yet to be identified.

FtsZ, the first identified prokaryotic cytoskeletal element, forms a filamentous ring structure located in the middle of the cell called the Z-ring that constricts during cell division, similar to the actin-myosin contractile ring in eukaryotes. The Z-ring is a highly dynamic structure that consists of numerous bundles of protofilaments that extend and shrink, although the mechanism behind Z-ring contraction and the number of protofilaments involved are unclear. FtsZ acts as an organizer protein and is required for cell division. It is the first component of the septum during cytokinesis, and it recruits all other known cell division proteins to the division site.

Despite this functional similarity to actin, FtsZ is homologous to eukaryal tubulin. Although comparison of the primary structures of FtsZ and tubulin reveal a weak relationship, their 3-dimensional structures are remarkably similar. Furthermore, like tubulin, monomeric FtsZ is bound to GTP and polymerizes with other FtsZ monomers with the hydrolysis of GTP in a mechanism similar to tubulin dimerization. Since FtsZ is essential for cell division in bacteria, this protein is a target for the design of new antibiotics.

MreB

MreB is a bacterial protein believed to be analogous to eukaryal actin. MreB and actin have a weak primary structure match, but are very similar in terms of 3-D structure and filament polymerization.

Almost all non-spherical bacteria rely on MreB to determine their shape. MreB assembles into a helical network of filamentous structures just under the cytoplasmic membrane, covering the whole length of the cell. MreB determines cell shape by mediating the position and activity of enzymes that synthesize peptidoglycan and by acting as a rigid filament under the cell membrane that exerts outward pressure to sculpt and bolster the cell. MreB condenses from its normal helical network and forms a tight ring at the septum in *Caulobacter crescentus* right before cell division, a mechanism that is believed to help locate its off-centre septum. MreB is also important for polarity determination in polar bacteria, as it is responsible for the correct positioning of at least four different polar proteins in *C. crescentus*.

Crescentin

Crescentin (encoded by *creS* gene) is an analogue of eukaryotic intermediate filaments (IFs). Unlike the other analogous relationships discussed here, crescentin has a rather large primary homology with IF proteins in addition to three-dimensional similarity - the sequence of *creS* has a 25% identity match and 40% similarity to cytokeratin 19 and a 24% identity match and 40% similarity to nuclear lamin A. Furthermore, crescentin filaments are roughly 10 nm in diameter and thus fall within diameter range for eukaryal IFs (8-15 nm). Crescentin forms a continuous filament from pole to pole alongside the inner, concave side of the crescent-shaped bacterium *Caulobacter crescentus*. Both MreB and crescentin are necessary for *C. crescentus* to exist in its characteristic shape; it is believed that MreB molds the cell into a rod shape and crescentin bends this shape into a crescent.

ParM and SopA

ParM is a cytoskeletal element that possesses a similar structure to actin, although it behaves functionally like tubulin. Further, it polymerizes bidirectionally and it exhibits dynamic instability, which are both behaviors characteristic of tubulin polymerization. It forms a system with ParR and *parC* that is responsible for R1 plasmid separation. ParM affixes to ParR, a DNA-binding protein that specifically binds to 10 direct repeats in the *parC* region on the R1 plasmid. This binding occurs on both ends of the ParM filament. This filament is then extended, separating the plasmids. The system is analogous to eukaryotic chromosome segregation as ParM acts like eukaryotic tubulin in the mitotic spindle, ParR acts like the kinetochore complex, and *parC* acts like the centromere of the chromosome. F plasmid segregation occurs in a similar system where SopA acts as the cytoskeletal filament and SopB binds to the *sopC* sequence in the F plasmid, like the kinetochore and centromere respectively.

MinCDE System

The MinCDE system is a filament system that properly positions the septum in the middle of the cell in *Escherichia coli*. According to Shih et al., MinC inhibits the formation of the septum by prohibiting the polymerization of the Z-ring. MinC, MinD, and MinE form a helix structure that winds around the cell and is bound to the membrane by MinD. The MinCDE helix occupies a pole and terminates in a filamentous structure called the E-ring made of MinE at the middle-most edge of the polar zone.

From this configuration, the E-ring will contract and move toward that pole, disassembling the MinCDE helix as it moves along. Concomitantly, the disassembled fragments will reassemble at the opposite polar end, reforming the MinCDE coil on the opposite pole while the current MinCDE helix is broken down. This process then repeats, with the MinCDE helix oscillating from pole to pole. This oscillation occurs repeatedly during the cell cycle, thereby keeping MinC (and its septum inhibiting effect) at a lower time-averaged concentration at the middle of the cell than at the ends of the cell.

The dynamic behavior of the Min proteins has been reconstituted in vitro using an artificial lipid bilayer as mimic for the cell membrane. MinE and MinD self-organized into parallel and spiral protein waves by a reaction-diffusion like mechanism.

Microtrabeculae

A fourth eukaryotic cytoskeletal element, *microtrabeculae*, was proposed by Keith Porter based on images obtained from high-voltage electron microscopy of whole cells in the 1970s. The images showed short, filamentous structures of unknown molecular composition associated with known cytoplasmic structures. Porter proposed that this microtrabecular structure represented a novel filamentous network distinct from microtubules, filamentous actin, or intermediate filaments. It is now generally accepted that microtrabeculae are nothing more than an artifact of certain types of fixation treatment, although we have yet to fully understand the complexity of the cell's cytoskeleton.

Cilia and Flagella

A cilium (plural *cilia*) is an organelle found in eukaryotic cells. Cilia are slender protuberances that project from the much larger cell body.

There are two types of cilia: *motile* cilia and *non-motile*, or primary cilia, which typically serve as sensory organelles. In eukaryotes, cilia and flagella together make up a group of organelles known as undulipodia. Eukaryotic cilia are structurally identical to Eukaryotic flagella, although distinctions are sometimes made according to function and/or length.

Types and Distribution

In humans, *primary* cilia are found on nearly every cell in the body. Cilia are rare in most plants, occurring most notably in cycads.

Ciliates are microscopic organisms that possess *motile* cilia exclusively and use them for either locomotion or to simply move liquid over their surface.

Larger eukaryotes, such as mammals, have *motile* cilia as well. Motile cilia are usually present on a cell's surface in large numbers and beat in coordinated waves. In humans, for example, motile cilia are found in the lining of the trachea (windpipe), where they sweep mucus and dirt out of the lungs. In female mammals, the beating of cilia in the Fallopian tubes moves the ovum from the ovary to the uterus.

In comparison to *motile* cilia, non-motile (or *primary*) cilia usually occur one per cell; nearly all mammalian cells have a single non-motile *primary cilium*. In addition, examples of specialized primary cilia can be found in human sensory organs such as the eye and the nose. The outer segment of the rod photoreceptor cell in the human eye is connected to its cell body with a specialized non-motile cilium. The dendritic knob of the olfactory neuron, where the odorant receptors are located, also contains non-motile cilia (about 10 cilia per dendritic knob).

Although the primary cilium was discovered in 1898, it was largely ignored for a century. Only recently has great progress been made in understanding the function of the primary cilium. Until the 1990s, the prevailing view of the primary cilium was that it was merely a vestigial organelle, without important function. Recent findings regarding its physiological roles in chemical sensation, signal transduction, and control of cell growth, have led scientists to acknowledge its importance in cell function, with the discovery of its role in diseases not previously recognized to involve the dysgenesis and dysfunction of cilia, such as polycystic kidney disease congenital heart disease, and an emerging group of genetic ciliopathies. The primary cilium is now known to play an important role in the function of many human organs. The current scientific understanding of primary cilia views them as "sensory cellular antennae that coordinate a large number of cellular signalling pathways, sometimes coupling the signalling to ciliary motility or alternatively to cell division and differentiation."

Structure, Assembly and Maintenance, and Function

"Inside cilia and flagella is a microtubule-based cytoskeleton called the axoneme. The axoneme of primary cilia typically has a ring of nine outer microtubule doublets (called a 9+0 axoneme), and the axoneme of a motile cilium has two central microtubule singlets in addition to the nine outer doublets (called a 9+2 axoneme). The axonemal cytoskeleton acts as a scaffolding for various protein complexes and provides binding sites for molecular motor proteins such as kinesin II, that help carry proteins up and down the microtubules."

The building blocks of the cilia such as tubulins and other partially assembled axonemal proteins are added to the ciliary tips which point away from the cell body. In most species bi-directional motility called intraflagellar transport or IFT plays an essential role to move these building materials from the cell body to the assembly site. IFT also carries the disassembled material to be recycled from the ciliary tip back to the cell body. By regulating the equilibrium between these two IFT processes, the length of cilia can be maintained dynamically. The disassembly of the cilia requires the action of the protein kinase Aurora A.

Exceptions where IFT is not present include *Plasmodium falciparum* which is one of the species of *Plasmodium* that cause malaria in humans. In this parasite, cilia assemble in the cytoplasm.

At the base of the cilium where the cilia attaches to the cell body is the microtubule organizing centre, the basal body. Some basal body proteins as CEP164, ODF2 and CEP170, regulate the formation and the stability of the cilium. A transition zone between the basal body and the axoneme "serves as a docking station for intraflagellar transport and motor proteins."

"In effect, the [cilium] is a nanomachine composed of perhaps over 600 proteins in molecular complexes, many of which also function independently as nanomachines."

Sensing the Extracellular Environment

"Some epithelial cells are ciliated, and they commonly exist as a sheet of polarised cells forming a tube or tubule with cilia projecting into the lumen." Primary cilia on epithelial cells provide chemosensation, thermosensation and mechanosensation of the extracellular environment by playing "a sensory role mediating specific signalling cues, including soluble factors in the external cell environment, a secretory role in which a soluble protein is released to have an effect downstream of the fluid flow, and mediation of fluid flow if the cilia are motile."

Cilium-related Disease

Ciliary defects can lead to a number of human diseases. Genetic mutations compromising the proper functioning of cilia, ciliopathies, can cause chronic disorders such as primary ciliary dyskinesia (PCD), nephronophthisis or Senior-Loken syndrome. In addition, a defect of the primary cilium in the renal tube cells can lead to polycystic kidney disease (PKD).

In another genetic disorder called Bardet-Biedl syndrome (BBS), the mutant gene products are the components in the basal body and cilia.

Lack of functional cilia in female Fallopian tubes can cause ectopic pregnancy. A fertilized ovum may not reach the uterus if the cilia are unable to move it there. In such a case, the ovum will implant in the Fallopian tubes, causing a tubal pregnancy, the most common form of ectopic pregnancy.

Since the flagellum of human sperm is actually a modified cilium, ciliary dysfunction can also be responsible for male infertility.

Of interest, there is an association of primary ciliary dyskinesia with left-right anatomic abnormalities such as situs inversus (a combination of findings known as Kartagener's syndrome) and other heterotaxic defects. These left-right anatomic abnormalities can also result in congenital heart disease. In fact, it has been shown that proper cilial function is responsible for the normal left-right asymmetry in mammals.

Ciliopathy as an Origin for Many Multi-symptom Genetic Diseases

Recent findings in genetic research have suggested that a large number of genetic disorders, both genetic syndromes and genetic diseases, that were not previously related in the medical literature, may be, in fact, highly related in the root cause of the widely-varying set of medical symptoms that are clinically visible in the disorder. These have been grouped as an emerging class of diseases called ciliopathies. The underlying cause may be a dysfunctional molecular mechanism in the primary cilia structures, organelles which are present in many diverse cellular types throughout the human body.

Cilia defects adversely affect "numerous critical developmental signalling pathways" essential to cellular development and thus offer a plausible hypothesis for the often multi-symptom nature of a large set of syndromes and diseases. Known ciliopathies include primary ciliary dyskinesia, Bardet-Biedl syndrome, polycystic kidney and liver disease, nephronophthisis, Alstrom syndrome, Meckel-Gruber syndrome and some forms of retinal degeneration.

Mucociliary Clearance

Mucociliary clearance, also referred to as mucociliary apparatus or mucociliar clearance (MCC), derived from *mucus*, *cilia* (cilia of the tracheal surface epithelium in the respiratory tract) and *clearance* describes the self-clearing mechanism of the bronchi.

The main bronchi down to the alveoli are lined with a respiratory epithelium. On that, cilium is present, bearing hair-shaped structures on its surface (cilia). The cilia are surrounded by a thin fluid film of mucus. On top of that is a second viscous film of mucus, in which foreign particles and microorganisms get stuck. Within the thin fluid film of mucus the cilia act out movements coordinated in direction towards the pharynx. Thereby the viscous film of mucus including its freight is transported off in direction towards the mouth, where it is either swallowed or expelled via coughing.

Important for good mucociliary clearance are the number of cilia, their structure, activity, and coordinated movement. Optimum functionality of mucociliary clearance presuppose a temperature of 37 °C and an absolute humidity of 44 mg/dm^3 corresponding to a relative humidity of 100 %. Under the condition of insufficient temperature and humidity, after a short time the ciliary cells suspend their transport function. Under such circumstances, bacterial germinal colonization is facilitated. Pulmonary infections and damaging of the pulmonic tissues may be the consequence. High humidity enhances the functionality of the mucociliary clearance. Two methods of supporting mucociliary clearance are the active and the passive respiratory gas humidification, which in particular are applied in mechanical ventilation.

Kinocilium

A kinocilium is a special type of cilium on the apex of hair cells located in the sensory epithelium of the vertebrate inner ear.

Anatomy in Humans

Kinocilia on hair cells in the human ear are involved in the senses of movement and hearing. In the auditory system, each hair cell has a kinocilium that is important in the depolarization or hyperpolarization of the cell. In the vestibular system, movement causes displacement of the stereocilia and kinocilium, resulting in depolarization or hyperpolarization of the cell. The depolarization of the hair cells in both instances causes signal transduction via neurotransmitter release.

Auditory System

In the human ear, each hair cell has a single non-motile cilium-like structure known as the kinocilium. The movement of other stereocilia on the hair cell towards the kinocilium, as a result of [[endolymph [1]]] flow, will cause potassium channels on the cilia to open due to a pulling force that each stereocilia exerts on its neighbouring cell (usually from tallest to shortest in a bundle of stereocilia) and this leads to the depolarization of the hair cell.

This pattern of depolarization should not be confused with the more common depolarization which involves the influx of Na+ into the cell which K+ channels stay closed. Endolymph composition resembles that of the intracellular fluid (more K+ and less Na+) more closely compared to its counterpart, perilymph which resembles the extracellular fluid (more Na+ and less K+ compared to intracellular matrix). This depolarization will open voltage gated calcium channels. The influx of calcium then triggers the cell to release vesicles containing excitatory neurotransmitters into a synapse. The post-synaptic neurite then sends an action potential to the spiral ganglion. Cochlear hair cells have kinocilia that degenerate following maturation of the stereociliary bundle. It is believed that the kinocilium is important in delineating the anatomical asymmetry of stereocilia or in making some mechanical connections with the hair cells in which they persist.

Vestibular Apparatus

Kinocilia are present in the crista ampullaris of the semicircular ducts and the sensory maculae of the utricle and saccule. One kinocilium is the longest cilium located on the hair cell next to 40-70 sterocilia. During movement of the body, the hair cell is depolarized when the sterocilia move toward the kinocilium. The depolarization of the hair cell causes neurotransmitter to be released and an increase in firing frequency of cranial nerve VIII. When the sterocilia tilt away from the kinocilium, the hair cell is hyperpolarized, decreasing the amount of neurotransmitter released, which decreases the firing frequency of cranial nerve VIII.

Anatomy in Fish and Frogs

The apical surface of a sensory fish hair cell usually has numerous stereocilia and a single, much longer kinocilium. Deflection of the stereocilia toward or away from the kinocilium causes an increase or decrease in the firing rate of the sensory neuron innervating the hair cell at its basal surface.

Hair cells in fish and some frogs are used to detect water movements around their bodies. These hair cells are embedded in a jelly-like protrusion called cupula. The hair cells therefore can not be seen and do not appear on the surface of skin of fish and frogs.

Flagellum

A flagellum is a tail-like projection that protrudes from the cell body of certain prokaryotic and eukaryotic cells, and functions in locomotion. There are some notable differences between prokaryotic and eukaryotic flagella, such as protein composition, structure, and mechanism of propulsion. An example of a flagellated bacterium is the ulcer-causing *Helicobacter pylori*, which uses multiple flagella to propel itself through the mucus lining to reach the stomach epithelium. An example of a eukaryotic flagellated cell is the sperm cell, which uses its flagellum to propel itself through the female reproductive tract. Eukaryotic flagella are structurally identical to eukaryotic cilia, although distinctions are sometimes made according to function and/ or length.

The word *flagellum* is the Latin word for whip.

Types

Three types of flagella have so far been distinguished; bacterial, archaeal and eukaryotic.

The main differences among these three types are summarized below:

- Bacterial flagella are helical filaments that rotate like screws. They provide two of several kinds of bacterial motility.
- Archaeal flagella are superficially similar to bacterial flagella, but are different in many details and considered non-homologous.
- Eukaryotic flagella - those of animal, plant, and protist cells - are complex cellular projections that lash back and forth.

Sometimes eukaryotic flagella are called *cilia* or *undulipodia* to emphasize their distinctiveness.

Bacterial

The bacterial flagellum is made up of the protein flagellin. Its shape is a 20 nanometer-thick hollow tube. It is helical and has a sharp bend just outside the outer membrane; this "hook" allows the helix to point directly away from the cell. A shaft runs between the hook and the basal body, passing through protein rings in the cell's membrane that act as bearings. Gram-positive organisms have 2 of these basal body rings, one in the peptidoglycan layer and one in the plasma membrane. Gram-negative organisms have 4 such rings: the L ring associates with the lipopolysaccharides, the P ring associates with peptidoglycan layer, the M ring is embedded in the plasma membrane, and the S ring is directly attached to the plasma membrane. The filament ends with a capping protein.

The bacterial flagellum is driven by a rotary engine (the *Mot complex*) made up of protein, located at the flagellum's anchor point on the inner cell membrane. The engine is powered by proton motive force, i.e., by the flow of protons (hydrogen ions) across the bacterial cell membrane due to a concentration gradient set up by the cell's metabolism (in *Vibrio* species there are two kinds of flagella, lateral and polar, and some are driven by a sodium ion pump rather than a proton pump). The rotor transports protons across the membrane, and is turned in the process. The rotor alone can operate at 6,000 to 17,000 rpm, but with the flagellar filament attached usually only reaches 200 to 1000 rpm. The direction of rotation can be switched almost instantaneously, caused by a slight change in the position of a protein, FliG, in the rotor.

The cylindrical shape of flagella is suited to locomotion of microscopic organisms; these organisms operate at a low Reynolds number, where the viscosity of the surrounding water is much more important than its mass or inertia.

Flagella do not rotate at a constant speed but instead can increase or decrease their rotational speed in relation to the strength of the proton motive force. Flagellar rotation can move bacteria through liquid media at speeds of up to 60 cell lengths/second (sec). Although this is only about 0.00017 km/h (0.00011 mph), when comparing this speed with that of higher organisms in terms of number of lengths moved per second, it is extremely fast. By comparison, the cheetah, the fastest land animal, can sprint at 110 km/h (68 mph), which is approximately 25 body lengths/sec.

During flagellar assembly, components of the flagellum pass through the hollow cores of the basal body and the nascent filament. During assembly, protein components are added at the flagellar tip rather than at the base. *In vitro*, flagellar filaments assemble spontaneously in a solution containing purified flagellin as the sole protein.

The flagellar filament is the long helical screw that propels the bacterium when rotated by the motor, through the hook. In most bacteria that have been studied, including the Gram negative *Escherichia coli, Salmonella typhimurium, Caulobacter crescentus*, and *Vibrio alginolyticus*, the filament is made up of eleven protofilaments approximately parallel to the filament axis. Each protofilament is a series of tandem protein chains. However in *Campylobacter jejuni*, there are seven protofilaments.

The basal body has several traits in common with some types of secretory pores, such as the hollow rod-like "plug" in their centres extending out through the plasma membrane. Given the structural similarities between bacterial flagella and bacterial secretory systems, it is thought that bacterial flagella may have evolved from the type three secretion system; however, it is not known for certain whether these pores are derived from the bacterial flagella or the bacterial secretory system. Through use of their flagella, *E. coli* are able to move rapidly towards attractants and away from repellents. They do this by means of a biased random walk, with 'runs' and 'tumbles' brought about by rotating the flagellum counter-clockwise and clockwise respectively.

Flagella Arrangement Schemes

Different species of bacteria have different numbers and arrangements of flagella. Monotrichous bacteria have a single flagellum (e.g., *Vibrio cholerae*). Lophotrichous bacteria have multiple flagella located at the same spot on the bacteria's surfaces which act in concert to drive the bacteria in a single direction. In many cases, the bases of multiple flagella are surrounded by a specialized region of the cell membrane, the so-called *polar membrane*. Amphitrichous bacteria have a single flagellum on each of two opposite ends (only one flagellum operates at a time, allowing the bacteria to reverse course rapidly by switching which flagellum is active). Peritrichous bacteria have flagella projecting in all directions (e.g., *E. coli*).

In some bacteria, such as the larger forms of *Selenomonas*, the individual flagella are organized outside the cell body, helically twining about each other to form a thick structure called a "fascicle". Other bacteria, such as Spirochetes, have a specialized type of flagellum called an "axial filament" that is located in the periplasmic space, the rotation of which causes the entire bacterium to move forward in a corkscrew-like motion.

Counterclockwise rotation of monotrichous polar flagella pushes the cell forward with the flagella trailing behind, much like a corkscrew moving inside cork. Indeed water in the microscopic scale is highly viscous, very different from our daily experience of water. The flagella are left-handed helices, and bundle and rotate together only when rotating counterclockwise. When some of the rotors reverse direction, the flagella unwind and the cell starts "tumbling". It has also been suggested that even if all flagella would rotate clockwise, they will not form a bundle, due to geometrical as well as hydrodynamical reasons.

Such "tumbling" may happen occasionally, leading to the cell seemingly thrashing about in place, resulting in the reorientation of the cell. The clockwise rotation of a flagellum is suppressed by chemical compounds favourable to the cell (e.g. food), but the motor is highly adaptive to this. Therefore, when moving in a favourable direction, the concentration of the chemical attractant increases and "tumbles" are continually suppressed; however, when the cell's direction of motion is unfavourable (e.g., away from a chemical attractant), tumbles are no longer suppressed and occur much more often, with the chance that the cell will be thus reoriented in the correct direction.

In some *Vibrio* spp. (particularly *Vibrio parahemolyticus*) and related proteobacteria such as *Aeromonas*, two flagellar systems co-exist, using different sets of genes and different ion gradients for energy. The polar flagella are constitutively expressed and provide motility in bulk fluid, while the lateral flagella are expressed when the polar flagella meet too much resistance to turn. These provide swarming motility on surfaces or in viscous fluids.

Archaeal

The archaeal flagellum is superficially similar to the bacterial (or eubacterial) flagellum; in the 1980s they were thought to be homologous on the basis of gross morphology and behavior. Both flagella consist of filaments extending outside of the cell, and rotate to propel the cell. Archaeal flagella have a unique structure which lacks a central channel. Similar to bacterial type IV pilins, the component flagellins are made with class 3 signal peptides and they are processed by a type IV prepilin peptidase-like enzyme. The archaeal flagellins are typically modified by the addition of N-linked glycans which are necessary for proper assembly and/or function.

Discoveries in the 1990s revealed numerous detailed differences between the archaeal and bacterial flagella; these include:

- Bacterial flagella are motorized by a flow of H^+ ions (or occasionally Na^+ ions); archaeal flagella are almost certainly powered by ATP. The torque-generating motor that powers rotation of the archaeal flagellum has not been identified.

- While bacterial cells often have many flagellar filaments, each of which rotates independently, the archaeal flagellum is composed of a bundle of many filaments that rotate as a single assembly.
- Bacterial flagella grow by the addition of flagellin subunits at the tip; archaeal flagella grow by the addition of subunits to the base.
- Bacterial flagella are thicker than archaeal flagella, and the bacterial filament has a large enough hollow "tube" inside that the flagellin subunits can flow up the inside of the filament and get added at the tip; the archaeal flagellum is too thin to allow this.
- Many components of bacterial flagella share sequence similarity to components of the type III secretion systems, but the components of bacterial and archaeal flagella share no sequence similarity. Instead, some components of archaeal flagella share sequence and morphological similarity with components of type IV pili, which are assembled through the action of type II secretion systems (the nomenclature of pili and protein secretion systems is not consistent).

These differences could mean that the bacterial and archaeal flagella could be a classic case of biological analogy, or convergent evolution, rather than homology. However, in comparison to the decades of well-publicized study of bacterial flagella (e.g. by Berg), archaeal flagella have only recently begun to get serious scientific attention.

Therefore, many assume erroneously that there is only one basic kind of prokaryotic flagellum, and that archaeal flagella are homologous to it. For example, Cavalier-Smith (2002) is aware of the differences between archaeal and bacterial flagellins, but retains the misconception that the basal bodies are homologous.

Eukaryotic

Along with cilia, flagella make up a group of organelles known as undulipodia.

Structure

A eukaryotic flagellum is a bundle of nine fused pairs of microtubule *doublets* surrounding two central single microtubules. The so-called *"9+2"* structure is characteristic of the core of the eukaryotic flagellum called an *axoneme.*

At the base of a eukaryotic flagellum is a basal body, "blepharoplast" or kinetosome, which is the microtubule organizing centre (MTOC) for flagellar microtubules and is about 500 nanometers long. Basal bodies are structurally identical to centrioles. The flagellum is encased within the cell's plasma membrane, so that the interior of the flagellum is accessible to the cell's cytoplasm.

Mechanism

Each of the outer 9 doublet microtubules extends a pair of dynein arms (an "inner" and an "outer" arm) to the adjacent microtubule; these dynein arms are responsible for flagellar beating, as the force produced by the arms causes the microtubule doublets to slide against each other and the flagellum as a whole to bend. These dynein arms produce force through ATP hydrolysis. The flagellar axoneme also contains radial spokes, polypeptide complexes extending from each of the outer 9 microtubule doublets towards the central pair, with the "head" of the spoke facing inwards. The radial spoke is thought to be involved in the regulation of flagellar motion, although its exact function and method of action are not yet understood.

Flagella vs Cilia

Though eukaryotic flagella and motile cilia are ultrastructurally identical, the beating pattern of the two organelles can be different. In the case of flagella (e.g. the tail of a sperm) the motion is propeller-like. In contrast, beating of motile cilia consists of coordinated back-and-forth cycling of many cilia on the cell surface. Thus, flagella serve for the propulsion of single cells (e.g. swimming of protozoa and spermatozoa), and motile cilia for the transport of fluids (e.g. transport of mucus by stationary ciliated cells in the trachea). However, cilia are also used for locomotion (through liquids) in organisms such as *Paramecium*.

Intraflagellar Transport

Intraflagellar transport (IFT), the process by which axonemal subunits, transmembrane receptors, and other proteins are moved up and down the length of the flagellum, is essential for proper functioning of the flagellum, in both motility and signal transduction.

Evolution of Flagella and Debate

Based on similarity in structure and partial similarity in amino acid sequence, it is generally accepted among scientists that the eukaryotic flagellum and cilium have evolved from the cytoskeleton, while the eubacterial flagellum has evolved either from the type III secretion system or from a more ancient secretion system from which the type III secretion system has evolved as well. The archaeal flagellum has probably evolved from the type IV pili. More details appear under Evolution of flagella. In his 1996 book *Darwin's Black Box*, intelligent design proponent Michael Behe, under funding from the Templeton Foundation, cited the bacterial flagellum as an example of an irreducibly complex structure that could not have evolved through naturalistic means. Behe argued that the flagellum becomes useless if any one of its constituent parts is removed, and thus could not have arisen through numerous, successive, slight modifications; therefore, it is hopelessly improbable that the proteins making up the flagellar motor could have come together all at once, by chance. Mark Perakh explained that while Behe popularized the idea, biologist Hermann J. Müller had already explored it (under the slightly different name of "interlocking complexity") and more than a decade before Behe's book the same idea was explored by A. Graham Cairns-Smith, but neither claimed that "irreducible complexity" was a "marker" of a supernatural design.

While Behe discussed the immune system and the blood clotting cascade in greater detail, the bacterial flagellum has become a "poster child" for intelligent design proponents and other creationists, as during the Kitzmiller v. Dover Area School District trial of 2005. It is one of two identified rotary structures found in nature (the other being ATP synthase) and it is billions of years older than Behe's other two examples, which exist in many homologous forms, simplifying the explanation of their origin.

Evolutionary pathways supported by the Theory of Natural Selection and Evolution have since been identified for the bacterial flagellum; thus, undermining Behe's argument. In addition, the Type three secretion system, a molecular syringe which bacteria use to inject toxins into other cells, appears to be a simplified sub-set of the bacterial flagellum's components, meaning that it is much less likely to be irreducibly complex in the way that the bacterial flagellum could have in fact evolved from the type three secretion system. Exaptation explains how systems with multiple parts can evolve through natural means.

Ciliopathy

A ciliopathy is a genetic disorder of the cellular cilia or the cilia anchoring structures, the basal bodies.

History

Although non-motile or primary cilia were first described in 1898, "cell biologists largely ignored them." But "microscopists continued to document their presence in the cells of most vertebrate organisms." The "primary cilium was long considered—with few exceptions—to be a largely useless evolutionary vestige, a vestigial organelle. Recent research has revealed an initial understanding that cilia are essential to many of the body's organs". Many mammalian eukaryotic cells are ciliated with *primary cilia*. These primary cilia play important roles in chemosensation, mechanosensation, and thermosensation. Cilia may thus be "viewed as sensory cellular antennae that coordinate a large number of cellular signalling pathways, sometimes coupling the signalling to ciliary motility or alternatively to cell division and differentiation."

Recent advances in mammalian genetic research have facilitated the elucidation of a molecular basis for a number of dysfunctional mechanisms in both motile and primary cilia structures of the cell. "Numerous critical developmental signalling pathways" essential to cellular development have been discovered. These are principally but not exclusively found in the *non-motile* or primary cilia. A number of common observable characteristics of mammalian genetic disorders and diseases are caused by ciliary dysgenesis and dysfunction. Once identified, these characteristics thus describe a set of hallmarks of a ciliopathy.

Cilia have recently been implicated in a wide variety of human genetic diseases by "the discovery that numerous proteins involved in mammalian disease localize to the basal bodies and cilia." For example, in just a single area of human disease physiology, cystic renal disease, cilia-related genes and proteins have been identified to have causal effect in polycystic kidney disease, nephronophthisis, Senior-Loken syndrome type 5, orofaciodigital syndrome type 1 and Bardet-Biedl syndrome.

The Mechanism of Ciliary Function

"In effect, the [motile cilium] is a nanomachine composed of perhaps over 600 proteins in molecular complexes, many of which also function independently as nanomachines." Cilia "function as mechano- or chemosensors and as a cellular global positioning system to detect changes in the surrounding environment." For example, ciliary signalling plays a role in the initiation of cellular replacement after cell damage. In addition to this sensory role mediating specific signalling cues, cilia play "a secretory role in which a soluble protein is released to have an effect downstream of the fluid flow" in epithelial cells, and can of course mediate fluid flow directly in the case of motile cilia. Primary cilia in the retina play a role in transferring nourishment to the non-vascularized rod and cone cells from the vascularized cells several microns behind the surface of the retina.

Similar Genes Can Result in a Range of Different Diseases

"Just as different genes can contribute to similar diseases, so the same genes and families of genes can play a part in a range of different diseases."

For example, in just two of the diseases caused by malfunctioning cilia, Meckel-Gruber syndrome and Bardet-Biedl syndrome, patients who carry mutations in genes associated with both diseases "have unique symptoms that are not seen in either condition alone." The genes linked to the two different conditions "interact with each other during development." Systems biologists are endeavoring to define functional modules containing multiple genes and then look at disorders whose phenotypes fit into such modules.

A particular phenotype can overlap "considerably with several conditions (ciliopathies) in which primary cilia are also implicated in pathogenicity. One emerging aspect is the wide spectrum of ciliopathy gene mutations found within different diseases."

Ciliopathies

"The phenotypic parameters that define a ciliopathy may be used to both recognize the cellular basis of a number of genetic disorders and to facilitate the diagnosis and treatment of some diseases of unknown etiology".

Syndromes and Diseases with a Known Cilia-based Causal Mechanism

- Alstrom syndrome,
- Bardet-Biedl syndrome
- early embryonic death (some cases)
- hydrocephalus (some cases)
- Joubert syndrome
- Meckel-Gruber syndrome
- nephronophthisis
- orofaciodigital syndrome type 1
- polycystic kidney disease
- polycystic liver disease
- primary ciliary dyskinesia

- retinal degeneration (some forms)
- Senior-Loken syndrome.

Implied or Suspected Ciliopathies

- agenesis of the corpus callosum
- anencephaly
- breathing abnomalities
- cerebellar vermis hypoplasia
- Dandy-Walker malformation
- diabetes
- Ellis-van Creveld syndrome
- exencephaly
- eye movement abnormalities
- liver disease
- hypoplasia of the corpus callosum
- hypotonia
- reproductive sterility
- Jeune asphyxiating thoracic dystrophy
- Juvenile myoclonic epilepsy (JME)
- Kartagener syndrome
- Marden-Walker syndrome
- "mental retardation/developmental delay" or "other cognitive defects"
- obesity
- polydactyly
- posterior encephalocele
- respiratory dysfunction
- "recurrent respiratory infections"
- renal cystic disease
- retinitis pigmentosa (some forms)
- sensorineural deafness
- situs inversus/Isomerism
- spina bifida.

Clinical Symptoms and Ciliac Roles

A wide variety of symptoms are potential clinical features of ciliopathy.

- Chemosensation abnormalities,, typically via ciliated epithelial cellular dysfunction.
- Defective thermosensation or mechanosensation, often via ciliated epithelial cellular dysfunction.
- Cellular motility dysfunction
- Issues with displacement of extracellular fluid
- Paracrine signal transduction abnormalities.

In organisms of normal health, cilia are critical for:

- organism development
- homeostasis
- reproduction.

Centrioles and Basal Bodies

A centriole is a barrel-shaped cell structure found in most animal eukaryotic cells, though absent in higher plants and most fungi. The walls of each centriole are usually composed of nine triplets of microtubules (protein of the cytoskeleton). Deviations from this structure include *Drosophila melanogaster* embryos, with nine doublets, and *Caenorhabditis elegans* sperm cells and early embryos, with nine singlets. An associated pair of centrioles, arranged perpendicularly and surrounded by an amorphous mass of dense material (the pericentriolar material) constitutes the compound structure known as the centrosome.

Cell Division

Centrioles are involved in the organization of the mitotic spindle and in the completion of cytokinesis. Centrioles were previously thought to be required for the formation of a mitotic spindle in animal cells. However, more recent experiments have demonstrated that cells whose centrioles have been removed via laser ablation can still progress through the G_1 stage of interphase before centrioles can be synthesized later in a de novo fashion. Additionally, mutant flies lacking centrioles can develop almost normally, although the adult flies lack flagella and cilia, a lack that underscores the requirement of centrioles for the formation of these organelles.

Cellular Organization

Centrioles are a very important part of centrosomes, which are involved in organizing microtubules in the cytoplasm. The position of the centriole determines the position of the nucleus and plays a crucial role in the spatial arrangement of the cell.

Ciliogenesis

In organisms with flagella and cilia, the position of these organelles is determined by the mother centriole, which becomes the basal body. An inability of cells to use centrioles to make functional cilia and flagella has been linked to a number of genetic and developmental diseases. In particular, the inability of centrioles to properly migrate prior to ciliary assembly has recently been linked to Meckel-Gruber syndrome.

Animal Development

Proper orientation of cilia via centriole positioning toward the posterior of embryonic node cells is critical for establishing left–right asymmetry during mammalian development.

Centriole Duplication

Cells in G_0 and G_1 usually contain two complete centrioles. The older of the two centrioles in a pair is termed the *mother centriole*, whereas the younger is termed the *daughter centriole*. During the cell division cycle, a new centriole grows from the side of each of the existing "mother" centrioles. After centriole duplication, the two pairs of centrioles remain attached to each other in an orthogonal configuration until mitosis, when the mother and daughter centrioles separate in a manner dependent upon the enzyme separase.

The two centrioles in the centrosome are connected to each other by unidentified proteins. The mother centriole has radiating appendages at the distal end of its long axis and is attached to the daughter centriole at the other proximal end. Each daughter cell formed after cell division will inherit one of these pairs (one older and one newer centriole). Duplication of centrioles starts at the time of the G_1/S transition and ends before the onset of mitosis.

A basal body (also called a basal granule or kinetosome) is an organelle formed from a centriole, and a short cylindrical array of microtubules. It is found at the base of a eukaryotic undulipodium (cilium or flagellum) and serves as a nucleation site for the growth of the axoneme microtubules. Centrioles, from which basal bodies are derived, act as anchoring sites for proteins that in turn anchor microtubules within centrosomes, one type of microtubule organizing centre (MTOC). These microtubules provide structure and facilitate movement of vesicles and organelles within many eukaryotic cells. Basal bodies, however, are specifically the bases for cilia and flagella that extend out of the cell.

Basal bodies are derived from centrioles through a largely mysterious process. They are structurally the same, each containing a microtubule triplet 9*3 helocoidal configuration forming a hollow cylinder. Regulation of basal body production and spatial orientation is a function of the nucleotide-binding domain of ã-tubulin (Shang et al., 2005).

Flagellum(flagella) are basically attached to the cell membrane from a basal granule.

The Nucleus

In cell biology, the nucleus (pl. *nuclei*; from Latin *nucleus* or *nuculeus*, meaning kernel) is a membrane-enclosed organelle found in eukaryotic cells. It contains most of the cell's genetic material, organized as multiple long linear DNA molecules in complex with a large variety of proteins, such as histones, to form chromosomes. The genes within these chromosomes are the cell's nuclear genome. The function of the nucleus is to maintain the integrity of these genes and to control the activities of the cell by regulating gene expression — the nucleus is therefore the control centre of the cell. The main structures making up the nucleus are the nuclear envelope, a double membrane that encloses the entire organelle and separates its contents from the cellular cytoplasm, and the nuclear lamina, a meshwork within the nucleus that adds mechanical support, much like the cytoskeleton supports the cell as a whole. Because the nuclear membrane is impermeable to most molecules, nuclear pores are required to allow movement of molecules across the envelope. These pores cross both of the membranes, providing a channel that allows free movement of small molecules and ions. The movement of larger molecules such as proteins is carefully controlled, and requires active transport regulated by carrier proteins. Nuclear transport is crucial to cell function, as movement through the pores is required for both gene expression and chromosomal maintenance.

Although the interior of the nucleus does not contain any membrane-bound subcompartments, its contents are not uniform, and a number of *subnuclear bodies* exist, made up of unique proteins, RNA molecules, and particular parts of the chromosomes. The best known of these is the nucleolus, which is mainly involved in the assembly of ribosomes. After being produced in the nucleolus, ribosomes are exported to the cytoplasm where they translate mRNA.

History

The nucleus was the first organelle to be discovered. The probably oldest preserved drawing dates back to the early microscopist Antonie van Leeuwenhoek (1632 – 1723). He observed a "Lumen", the nucleus, in the red blood cells of salmon. Unlike mammalian red blood cells, those of other vertebrates still possess nuclei. The nucleus was also described by Franz Bauer in 1804 and in more detail in 1831 by Scottish botanist Robert Brown in a talk at the Linnean Society of London. Brown was studying orchids microscopically when he observed an opaque area, which he called the areola or nucleus, in the cells of the flower's outer layer. He did not suggest a potential function. In 1838 Matthias Schleiden proposed that the nucleus plays a role in generating cells, thus he introduced the name "Cytoblast" (cell builder). He believed that he had observed new cells assembling around "cytoblasts". Franz Meyen was a strong opponent of this view having already described cells multiplying by division and believing that many cells would have no nuclei. The idea that cells can be generated de novo, by the "cytoblast" or otherwise, contradicted work by Robert Remak (1852) and Rudolf Virchow (1855) who decisively propagated the new paradigm that cells are generated solely by cells ("Omnis cellula e cellula"). The function of the nucleus remained unclear.

Between 1876 and 1878 Oscar Hertwig published several studies on the fertilization of sea urchin eggs, showing that the nucleus of the sperm enters the oocyte and fuses with its nucleus. This was the first time it was suggested that an individual develops from a (single) nucleated cell. This was in contradiction to Ernst Haeckel's theory that the complete phylogeny of a species would be repeated during embryonic development, including generation of the first nucleated cell from a "Monerula", a structureless mass of primordial mucus ("Urschleim"). Therefore, the necessity of the sperm nucleus for fertilization was discussed for quite some time. However, Hertwig confirmed his observation in other animal groups, e.g. amphibians and molluscs. Eduard Strasburger produced the same results for plants (1884). This paved the way to assign the nucleus an important role in heredity. In 1873 August Weismann postulated the equivalence of the maternal and paternal germ *cells* for heredity. The function of the nucleus as carrier of genetic information became clear only later, after mitosis was discovered and the Mendelian rules were rediscovered at the beginning of the 20th century; the chromosome theory of heredity was developed.

Structures

The nucleus is the largest cellular organelle in animals. In mammalian cells, the average diameter of the nucleus is approximately 6 micrometers (ìm), which occupies about 10% of the total cell volume. The viscous liquid within it is called nucleoplasm, and is similar in composition to the cytosol found outside the nucleus. It appears as a dense, roughly spherical organelle.

Nuclear Envelope and Pores

The nuclear envelope otherwise known as nuclear membrane consists of two cellular membranes, an inner and an outer membrane, arranged parallel to one another and separated by 10 to 50 nanometers (nm). The nuclear envelope completely encloses the nucleus and separates the cell's genetic material from the surrounding cytoplasm, serving as a barrier to prevent macromolecules from diffusing freely between the nucleoplasm and the cytoplasm.

The outer nuclear membrane is continuous with the membrane of the rough endoplasmic reticulum (RER), and is similarly studded with ribosomes. The space between the membranes is called the perinuclear space and is continuous with the RER lumen.

Nuclear pores, which provide aqueous channels through the envelope, are composed of multiple proteins, collectively referred to as nucleoporins. The pores are about 125 million daltons in molecular weight and consist of around 50 (in yeast) to 100 proteins (in vertebrates). The pores are 100 nm in total diameter; however, the gap through which molecules freely diffuse is only about 9 nm wide, due to the presence of regulatory systems within the centre of the pore. This size allows the free passage of small water-soluble molecules while preventing larger molecules, such as nucleic acids and larger proteins, from inappropriately entering or exiting the nucleus. These large molecules must be actively transported into the nucleus instead. The nucleus of a typical mammalian cell will have about 3000 to 4000 pores throughout its envelope, each of which contains a donut-shaped, eightfold-symmetric ring-shaped structure at a position where the inner and outer membranes fuse.

Attached to the ring is a structure called the *nuclear basket* that extends into the nucleoplasm, and a series of filamentous extensions that reach into the cytoplasm. Both structures serve to mediate binding to nuclear transport proteins.

Most proteins, ribosomal subunits, and some RNAs are transported through the pore complexes in a process mediated by a family of transport factors known as karyopherins. Those karyopherins that mediate movement into the nucleus are also called importins, while those that mediate movement out of the nucleus are called exportins. Most karyopherins interact directly with their cargo, although some use adaptor proteins.

Steroid hormones such as cortisol and aldosterone, as well as other small lipid-soluble molecules involved in intercellular signalling can diffuse through the cell membrane and into the cytoplasm, where they bind nuclear receptor proteins that are trafficked into the nucleus. There they serve as transcription factors when bound to their ligand; in the absence of ligand many such receptors function as histone deacetylases that repress gene expression.

Nuclear Lamina

In animal cells, two networks of intermediate filaments provide the nucleus with mechanical support: the nuclear lamina forms an organized meshwork on the internal face of the envelope, while less organized support is provided on the cytosolic face of the envelope. Both systems provide structural support for the nuclear envelope and anchoring sites for chromosomes and nuclear pores.

The nuclear lamina is mostly composed of lamin proteins. Like all proteins, lamins are synthesized in the cytoplasm and later transported into the nucleus interior, where they are assembled before being incorporated into the existing network of nuclear lamina. Lamins are also found inside the nucleoplasm where they form another regular structure, known as the *nucleoplasmic veil*, that is visible using fluorescence microscopy. The actual function of the veil is not clear, although it is excluded from the nucleolus and is present during interphase. The lamin structures that make up the veil bind chromatin and disrupting their structure inhibits transcription of protein-coding genes. Like the components of other intermediate filaments, the lamin monomer contains an alpha-helical domain used by two monomers to coil around each other, forming a dimer structure called a coiled coil. Two of these dimer structures then join side by side, in an antiparallel arrangement, to form a tetramer called a *protofilament*. Eight of these protofilaments form a lateral arrangement that is twisted to form a ropelike *filament*. These filaments can be assembled or disassembled in a dynamic manner, meaning that changes in the length of the filament depend on the competing rates of filament addition and removal.

Mutations in lamin genes leading to defects in filament assembly are known as *laminopathies*. The most notable laminopathy is the family of diseases known as progeria, which causes the appearance of premature aging in its sufferers. The exact mechanism by which the associated biochemical changes give rise to the aged phenotype is not well understood.

Chromosomes

The cell nucleus contains the majority of the cell's genetic material, in the form of multiple linear DNA molecules organized into structures called chromosomes. During most of the cell cycle these are organized in a DNA-protein complex known as chromatin, and during cell division the chromatin can be seen to form the well defined chromosomes familiar from a karyotype. A small fraction of the cell's genes are located instead in the mitochondria.

There are two types of chromatin. Euchromatin is the less compact DNA form, and contains genes that are frequently expressed by the cell. The other type, heterochromatin, is the more compact form, and contains DNA that are infrequently transcribed. This structure is further categorized into *facultative* heterochromatin, consisting of genes that are organized as heterochromatin only in certain cell types or at certain stages of development, and *constitutive* heterochromatin that consists of chromosome structural components such as telomeres and centromeres. During interphase the chromatin organizes itself into discrete individual patches, called *chromosome territories.* Active genes, which are generally found in the euchromatic region of the chromosome, tend to be located towards the chromosome's territory boundary.

Antibodies to certain types of chromatin organization, particularly nucleosomes, have been associated with a number of autoimmune diseases, such as systemic lupus erythematosus. These are known as anti-nuclear antibodies (ANA) and have also been observed in concert with multiple sclerosis as part of general immune system dysfunction. As in the case of progeria, the role played by the antibodies in inducing the symptoms of autoimmune diseases is not obvious.

Nucleolus

The nucleolus is a discrete densely stained structure found in the nucleus. It is not surrounded by a membrane, and is sometimes called a *suborganelle.* It forms around tandem repeats of rDNA, DNA coding for ribosomal RNA (rRNA). These regions are called nucleolar organizer regions (NOR). The main roles of the nucleolus are to synthesize rRNA and assemble ribosomes.

The structural cohesion of the nucleolus depends on its activity, as ribosomal assembly in the nucleolus results in the transient association of nucleolar components, facilitating further ribosomal assembly, and hence further association. This model is supported by observations that inactivation of rDNA results in intermingling of nucleolar structures.

The first step in ribosomal assembly is transcription of the rDNA, by a protein called RNA polymerase I, forming a large pre-rRNA precursor. This is cleaved into the subunits 5.8S, 18S, and 28S rRNA. The transcription, post-transcriptional processing, and assembly of rRNA occurs in the nucleolus, aided by small nucleolar RNA (snoRNA) molecules, some of which are derived from spliced introns from messenger RNAs encoding genes related to ribosomal function. The assembled ribosomal subunits are the largest structures passed through the nuclear pores.

When observed under the electron microscope, the nucleolus can be seen to consist of three distinguishable regions: the innermost *fibrillar centres* (FCs), surrounded by the *dense fibrillar component* (DFC), which in turn is bordered by the *granular component* (GC). Transcription of the rDNA occurs either in the FC or at the FC-DFC boundary, and therefore when rDNA transcription in the cell is increased more FCs are detected. Most of the cleavage and modification of rRNAs occurs in the DFC, while the latter steps involving protein assembly onto the ribosomal subunits occur in the GC.

Other Subnuclear Bodies

Table: *Subnuclear structure sizes*

Structure name	*Structure diameter*
Cajal bodies	0.2–2.0 μm
PIKA	5 μmPML
bodies	0.2–1.0 μm
Paraspeckles	0.2–1.0 μm
Speckles	20–25 nm

Besides the nucleolus, the nucleus contains a number of other non-membrane delineated bodies. These include Cajal bodies, Gemini of coiled bodies, polymorphic interphase karyosomal association (PIKA), promyelocytic leukaemia (PML) bodies, paraspeckles and splicing speckles. Although little is known about a number of these domains, they are significant in that they show that the nucleoplasm is not uniform mixture, but rather contains organized functional subdomains.

Other subnuclear structures appear as part of abnormal disease processes. For example, the presence of small intranuclear rods have been reported in some cases of nemaline myopathy. This condition typically results from mutations in actin, and the rods themselves consist of mutant actin as well as other cytoskeletal proteins.

Cajal Bodies and Gems

A nucleus typically contains between 1 and 10 compact structures called Cajal bodies or coiled bodies (CB), whose diameter measures between 0.2 μm and 2.0 μm depending on the cell type and species. When seen under an electron microscope, they resemble balls of tangled thread and are dense foci of distribution for the protein coilin. CBs are involved in a number of different roles relating to RNA processing, specifically small nucleolar RNA (snoRNA) and small nuclear RNA (snRNA) maturation, and histone mRNA modification.

Similar to Cajal bodies are Gemini of coiled bodies, or gems, whose name is derived from the Gemini constellation in reference to their close "twin" relationship with CBs. Gems are similar in size and shape to CBs, and in fact are virtually indistinguishable under the microscope. Unlike CBs, gems do not contain small nuclear ribonucleoproteins (snRNPs), but do contain a protein called *survivor of motor neurons* (SMN) whose function relates to snRNP biogenesis. Gems are believed to assist CBs in snRNP biogenesis, though it has also been suggested from microscopy evidence that CBs and gems are different manifestations of the same structure.

PIKA and PTF Domains

PIKA domains, or polymorphic interphase karyosomal associations, were first described in microscopy studies in 1991. Their function was and remains unclear, though they were not thought to be associated with active DNA replication, transcription, or RNA processing. They have been found to often associate with discrete domains defined by dense localization of the transcription factor PTF, which promotes transcription of snRNA.

PML Bodies

Promyelocytic leukaemia bodies (PML bodies) are spherical bodies found scattered throughout the nucleoplasm, measuring around 0.2–1.0 μm. They are known by a number of other names, including nuclear domain 10 (ND10), Kremer bodies, and PML oncogenic domains. They are often seen in the nucleus in association with Cajal bodies and cleavage bodies. It has been suggested that they play a role in regulating transcription.

Paraspeckles

Discovered by Fox et al. in 2002, paraspeckles are irregularly shaped compartments in the nucleus' interchromatin space. First documented in HeLa cells, where there are generally 10–30 per nucleus, paraspeckles are now known to also exist in all human primary cells, transformed cell lines and tissue sections. Their name is derived from their distribution in the nucleus; the "para" is short for parallel and the "speckles" refers to the splicing speckles to which they are always in close proximity.

Paraspeckles are dynamic structures that are altered in response to changes in cellular metabolic activity. They are transcription dependent and in the absence of RNA Pol II transcription, the paraspeckle disappears and all of its associated protein components (PSP1, p54nrb, PSP2, CFI(m)68 and PSF) form a crescent shaped perinucleolar cap in the nucleolus. This phenomenon is demonstrated during the cell cycle. In the cell cycle, paraspeckles are present during interphase and during all of mitosis except for telophase. During telophase, when the two daughter nuclei are formed, there is no RNA Pol II transcription so the protein components instead form a perinucleolar cap.

Splicing Speckles

Sometimes referred to as *interchromatin granule clusters* or as *splicing-factor compartments*, speckles are rich in splicing snRNPs and other splicing proteins necessary for pre-mRNA processing. Because of a cell's changing requirements, the composition and location of these bodies changes according to mRNA transcription and regulation via phosphorylation of specific proteins.

Function

The main function of the cell nucleus is to control gene expression and mediate the replication of DNA during the cell cycle. The nucleus provides a site for genetic transcription that is segregated from the location of translation in the cytoplasm, allowing levels of gene regulation that are not available to prokaryotes.

Cell Compartmentalization

The nuclear envelope allows the nucleus to control its contents, and separate them from the rest of the cytoplasm where necessary. This is important for controlling processes on either side of the nuclear membrane. In some cases where a cytoplasmic process needs to be restricted, a key participant is removed to the nucleus, where it interacts with transcription factors to downregulate the production of certain enzymes in the pathway. This regulatory mechanism occurs in the case of glycolysis, a cellular pathway for breaking down glucose to produce energy. Hexokinase is an enzyme responsible for the first the step of glycolysis, forming glucose-6-phosphate from glucose. At high concentrations of fructose-6-phosphate, a molecule made later from glucose-6-phosphate, a regulator protein removes hexokinase to the nucleus, where it forms a transcriptional repressor complex with nuclear proteins to reduce the expression of genes involved in glycolysis.

In order to control which genes are being transcribed, the cell separates some transcription factor proteins responsible for regulating gene expression from physical access to the DNA until they are activated by other signalling pathways. This prevents even low levels of inappropriate gene expression. For example in the case of NF-κB-controlled genes, which are involved in most inflammatory responses, transcription is induced in response to a signal pathway such as that initiated by the signalling molecule TNF-α, binds to a cell membrane receptor, resulting in the recruitment of signalling proteins, and eventually activating the transcription factor NF-κB. A nuclear localisation signal on the NF-κB protein allows it to be transported through the nuclear pore and into the nucleus, where it stimulates the transcription of the target genes.

The compartmentalization allows the cell to prevent translation of unspliced mRNA. Eukaryotic mRNA contains introns that must be removed before being translated to produce functional proteins. The splicing is done inside the nucleus before the mRNA can be accessed by ribosomes for translation. Without the nucleus ribosomes would translate newly transcribed (unprocessed) mRNA resulting in misformed and nonfunctional proteins.

Gene Expression

Gene expression is the process by which information from a gene is used in the synthesis of a functional gene product. These products are often proteins, but in non-protein coding genes such as rRNA genes or tRNA genes, the product is a functional RNA. The process of gene expression is used by all known life - eukaryotes (including multicellular organisms), prokaryotes (bacteria and archaea) and viruses - to generate the macromolecular machinery for life. Several steps in the gene expression process may be modulated, including the transcription, RNA splicing, translation, and post-translational modification of a protein. Gene regulation gives the cell control over structure and function, and is the basis for cellular differentiation, morphogenesis and the versatility and adaptability of any organism. Gene regulation may also serve as a substrate for evolutionary change, since control of the timing, location, and amount of gene expression can have a profound effect on the functions (actions) of the gene in a cell or in a multicellular organism.

In genetics, gene expression is the most fundamental level at which genotype gives rise to the phenotype. The genetic code is "interpreted" by gene expression, and the properties of the expression products give rise to the organism's phenotype.

Mechanism

Transcription

The gene itself is typically a long stretch of DNA. It is a blueprint for the production of RNA. The production of RNA copies of the DNA is called transcription, and is performed by RNA polymerase, which adds one RNA nucleotide at a time to a growing RNA strand. This RNA is complementary to the template 3' '! 5' DNA strand, which is itself complementary to the coding 5' '! 3' DNA strand. Therefore, the resulting 5' → 3' RNA strand is identical to the coding DNA strand with the exception that thymines are replaced with uracils in the RNA. A coding DNA strand reading "ATG" is transcribed as "AUG" in RNA.

RNA Processing

Transcription of protein encoding genes creates a primary transcript of RNA at the place where the gene was located. This transcript can be altered before being translated, this is particularly common in eukaryotes. The most common RNA processing is splicing to remove introns. Introns are RNA segments which are not found in the mature RNA, although they can function as precursors, e.g. for snoRNAs, which are RNAs that direct modification of nucleotides in other RNAs. Introns are common in eukaryotic genes but rare in prokaryotes.

RNA processing, also known as post-transcriptional modification, can start during transcription, as is the case for splicing, where the spliceosome removes introns from newly formed RNA.

Extensive RNA processing may be an evolutionary advantage made possible by the nucleus of eukaryotes. In prokaryotes transcription and translation happen together whilst in eukaryotes the nuclear membrane separates the two processes giving time for RNA processing to occur.

Non-coding RNA Maturation

In most organisms non-coding genes (ncRNA) are transcribed as precursors which undergo further processing. In the case of ribosomal RNAs (rRNA), they are often transcribed as a pre-rRNA which contains one or more rRNAs, the pre-rRNA is cleaved and modified (22 -O-methylation and pseudouridine formation) at a specific sites by approximately 150 different small nucleolus-restricted RNA species, called small nucleolar RNAs(snoRNAs), which like snRNAs, snoRNAs associate with proteins, forming snoRNPs. In eukaryotes, in particular a snoRNP, called RNase MRP cleaves the 45S pre-rRNA into the 28S, 5.8S, and 18S rRNAs. The rRNA and RNA processing factors are form large aggregates called the nucleolus.

In the case of transfer RNA (tRNA), for example, the 5' sequence is removed by RNase P, whereas the 3' end is removed by the tRNase Z enzyme. In the case of micro RNA (miRNA), miRNAs are first transcribed as primary transcripts or pri-miRNA with a cap and poly-A tail and processed to short, 70-nucleotide stem-loop structures known as pre-miRNA in the cell nucleus by the enzymes Drosha and Pasha, after being exported, it is then processed to mature miRNAs in the cytoplasm by interaction with the endonuclease Dicer, which also initiates the formation of the RNA-induced silencing complex (RISC), composed of the Argonaute protein.

RNA Export

In eukaryotes most mature RNA must be exported to the cytoplasm from the nucleus. While some RNAs function in the nucleus, many RNAs are transported through the nuclear pores and into the cytosol. Notably this includes all RNA types involved in protein synthesis. In some cases RNAs are additionally transported to a specific part of the cytoplasm, such as a synapse; they are then towed by motor proteins that bind through linker proteins to specific sequences (called "zipcodes") on the RNA.

Translation

For some RNA (non-coding RNA) the mature RNA is the finished gene product. In the case of messenger RNA (mRNA) the RNA is an information carrier coding for the synthesis of one or more proteins. mRNA carrying a single protein sequence (common in eukaryotes) is monocistronic whilst mRNA carrying multiple protein sequences (common in prokaryotes) is known as polycistronic.

Each triplet of nucleotides of the coding regions of a messenger RNA corresponds to a binding site for a transfer RNA. Transfer RNAs carry amino acids, and these are chained together by the ribosome. The ribosome helps transfer RNA to bind to messenger RNA and takes the amino acid from each transfer RNA and makes a structure-less protein out of it. In prokaryotes translation generally occurs at the point of transcription, often using a messenger RNA which is still in the process of being created. In eukaryotes translation can occur in a variety of regions of the cell depending on where the protein being written is supposed to be. Major locations are the cytoplasm for soluble cytoplasmic proteins and the endoplasmic reticulum for proteins which are for export from the cell or insertion into a cell membrane. Proteins which are supposed to be expressed at the endoplasmic reticulum are recognised part-way through the translation process. This is governed by the signal recognition particle - a protein which binds to the ribosome and directs it to the endoplasmic reticulum when it finds a signal sequence on the growing (nascent) amino acid chain.

Folding

Enzymes called chaperones assist the newly formed protein to attain (fold into) the 3-dimensional structure it needs to function. Similarly, RNA chaperones help RNAs attain their functional shapes. Assisting protein folding is one of the main roles of the endoplasmic reticulum in eukaryotes.

Protein Transport

Many proteins are destined for other parts of the cell than the cytosol and a wide range of signalling sequences are used to direct proteins to where they are supposed to be. In prokaryotes this is normally a simple process due to limited compartmentalisation of the cell. However in eukaryotes there is a great variety of different targeting processes to ensure the protein arrives at the correct organelle.

Not all proteins remain within the cell and many are exported, for example digestive enzymes, hormones and extracellular matrix proteins. In eukaryotes the export pathway is well developed and the main mechanism for the export of these proteins is translocation to the endoplasmatic reticulum, followed by transport via the Golgi apparatus.

Measurement

Measuring gene expression is an important part of many life sciences - the ability to quantify the level at which a particular gene is expressed within a cell, tissue or organism can give a huge amount of information. For example measuring gene expression can:

- Identify viral infection of a cell (viral protein expression)
- Determine an individual's susceptibility to cancer (oncogene expression)
- Find if a bacterium is resistant to penicillin (beta-lactamase expression).

Similarly the analysis of the location of expression protein is a powerful tool and this can be done on an organism or cellular scale.

Investigation of localisation is particularly important for study of development in multicellular organisms and as an indicator of protein function in single cells. Ideally measurement of expression is done by detecting the final gene product (for many genes this is the protein) however it is often easier to detect one of the precursors, typically mRNA, and infer gene expression level.

mRNA Quantification

Levels of mRNA can be quantitatively measured by Northern blotting which gives size and sequence information about the mRNA molecules. A sample of RNA is separated on an agarose gel and hybridized to a radio-labeled RNA probe that is complementary to the target sequence. The radio-labeled RNA is then detected by an autoradiograph. The main problems with Northern blotting stem from the use of radioactive reagents (which make the procedure time consuming and potentially dangerous) and lower quality quantification than more modern methods (due to the fact that quantification is done by measuring band strength in an image of a gel). Northern blotting is, however, still widely used as the additional mRNA size information allows the discrimination of alternately spliced transcripts.

A more modern low-throughput approach for measuring mRNA abundance is reverse transcription quantitative polymerase chain reaction (RT-PCR followed with qPCR). RT-PCR first generates a DNA template from the mRNA by reverse transcription, which is called cDNA. This cDNA template is then used for qPCR where the change in fluorescence of a probe changes as the DNA amplification process progresses. With a carefully constructed standard curve qPCR can produce an absolute measurement such as number of copies of mRNA, typically in units of copies per nanolitre of homogenized tissue or copies per cell. qPCR is very sensitive (detection of a single mRNA molecule is possible), but can be expensive due to the fluorescent probes required.

Northern blots and RT-qPCR are good for detecting whether a single gene is being expressed, but it quickly becomes impractical if many genes within the sample are being studied. Using DNA microarrays transcript levels for many genes at once (expression profiling) can be measured. Recent advances in microarray technology allow for the quantification, on a single array, of transcript levels for every known gene in several organism's genomes, including humans.

Alternatively "tag based" technologies like Serial analysis of gene expression (SAGE), which can provide a relative measure of the cellular concentration of different messenger RNAs, can be used. The great advantage of tag-based methods is the "open architecture", allowing for the exact measurement of any transcript, with a known or unknown sequence.

Protein Quantification

For genes encoding proteins the expression level can be directly assessed by a number of means with some clear analogies to the techniques for mRNA quantification.

The most commonly used method is to perform a Western blot against the protein of interest - this gives information on the size of the protein in addition to its identity. A sample (often cellular lysate) is separated on a polyacrylamide gel, transferred to a membrane and then probed with an antibody to the protein of interest. The antibody can either be conjugated to a fluorophore or to horseradish peroxidase for imaging and/or quantification. The gel-based nature of this assay makes quantification less accurate but it has the advantage of being able to identify later modifications to the protein, for example proteolysis or ubiquitination, from changes in size.

By replacing the gene with a new version fused a green fluorescent protein (or similar) marker expression may be directly quantified in live cells. This is done by imaging using a fluorescence microscope. It is very difficult to clone a GFP-fused protein into its native location in the genome without affecting expression levels so this method often cannot be used to measure endogenous gene expression. It is, however, widely used to measure the expression of a gene artificially introduced into the cell, for example via an expression vector. It is important to note that by fusing a target protein to a fluorescent reporter the protein's behavior, including its cellular localization and expression level, can be significantly changed.

The enzyme-linked immunosorbent assay works by using antibodies immobilised on a microtiter plate to capture proteins of interest from samples added to the well. Using a detection antibody conjugated to an enzyme or fluorophore the quantity of bound protein can be accurately measured by fluorometric or colourimetric detection. The detection process is very similar to that of a Western blot, but by avoiding the gel steps more accurate quantification can be achieved.

Localisation

Analysis of expression is not limited to only quantification; localisation can also be determined. mRNA can be detected with a suitably labelled complementary mRNA strand and protein can be detected via labelled antibodies. The probed sample is then observed by microscopy to identify where the mRNA or protein is.

Regulation of Gene Expression

Regulation of gene expression refers to the control of the amount and timing of appearance of the functional product of a gene. Control of expression is vital to allow a cell to produce the gene products it needs when it needs them; in turn this gives cells the flexibility to adapt to a variable environment, external signals, damage to the cell, etc. Some simple examples of where gene expression is important are:

- Control of Insulin expression so it gives a signal for blood glucose regulation
- X chromosome inactivation in female mammals to prevent an "overdose" of the genes it contains.
- Cyclin expression levels control progression through the eukaryotic cell cycle.

More generally gene regulation gives the cell control over all structure and function, and is the basis for cellular differentiation, morphogenesis and the versatility and adaptability of any organism. Any step of gene expression may be modulated, from the DNA-RNA transcription step to post-translational modification of a protein. The stability of the final gene product, whether it is RNA or protein, also contributes to the expression level of the gene - an unstable product results in a low expression level. In general gene expression is regulated through changes in the number and type of interactions between molecules that collectively influence transcription of DNA and translation of RNA.

Transcriptional Regulation

Regulation of transcription can be broken down into three main routes of influence; genetic (direct interaction of a control factor with the gene), modulation (interaction of a control factor with the transcription machinery) and epigenetic (non-sequence changes in DNA structure which influence transcription).

Direct interaction with DNA is the simplest and most direct method a protein can change transcription levels and genes often have several protein binding sites around the coding region with the specific function of regulating transcription. There are many classes of regulatory DNA binding sites known as enhancers, insulators, repressors and silencers. The mechanisms for regulating transcription are very varied, from blocking key binding sites on the DNA for RNA polymerase to acting as an activator and promoting transcription by assisting RNA polymerase binding.

The activity of transcription factors is further modulated by intracellular signals causing protein post-translational modification including phosphorylated, acetylated, or glycosylated. These changes influence a transcription factor's ability to bind, directly or indirectly, to promoter DNA, to recruit RNA polymerase, or to favour elongation of a newly synthetized RNA molecule.

The nuclear membrane in eukaryotes allows further regulation of transcription factors by the duration of their presence in the nucleus which is regulated by reversible changes in their structure and by binding of other proteins. Environmental stimuli or endocrine signals may cause modification of regulatory proteins eliciting cascades of intracellular signals, which result in regulation of gene expression.

More recently it has become apparent that there is a huge influence of non-DNA-sequence specific effects on translation. These effects are referred to as epigenetic and involve the higher order structure of DNA, non-sequence specific DNA binding proteins and chemical modification of DNA. In general epigenetic effects alter the accessibility of DNA to proteins and so modulate transcription.

DNA methylation is a widespread mechanism for epigenetic influence on gene expression and is seen in bacteria and eukaryotes and has roles in heritable transcription silencing and transcription regulation. In eukaryotes the structure of chromatin, controlled by the histone code, regulates access to DNA with significant impacts on the expression of genes in euchromatin and heterochromatin areas.

Post-transcriptional Regulation

In eukaryotes, where export of RNA is required before translation is possible, nuclear export is thought to provide additional control over gene expression. All transport in and out of the nucleus is via the nuclear pore and transport is controlled by a wide range of importin and exportin proteins.

Expression of a gene coding for a protein is only possible if the messenger RNA carrying the code survives long enough to be translated. In a typical cell an RNA molecule is only stable if specifically protected from degradation. RNA degradation has particular importance in regulation of expression in eukaryotic cells where mRNA has to travel significant distances before being translated. In eukaryotes RNA is stabilised by certain post-transcriptional modifications, particularly the 5' cap and poly-adenylated tail.

Intentional degradation of mRNA is used not just as a defence mechanism from foreign RNA (normally from viruses) but also as a route of mRNA *destabilisation*. If an mRNA molecule has a complementary sequence to a small interfering RNA then it is targeted for destruction via the RNA interference pathway.

Translational Regulation

Direct regulation of translation is less prevalent than control of transcription or mRNA stability but is occasionally used. Inhibition of protein translation is a major target for toxins and antibiotics in order to kill a cell by overriding its normal gene expression control. Protein synthesis inhibitors include the antibiotic neomycin and the toxin ricin.

Protein Degradation

Once protein synthesis is complete the level of expression of that protein can be reduced by protein degradation. There are major protein degradation pathways in all prokaryotes and eukaryotes of which the proteasome is a common component. An unneeded or damaged protein is often labelled for degradation by addition of ubiquitin.

Expression System

An expression system is a system specifically designed for the production of a gene product of choice. This is normally a protein although may also be RNA, such as tRNA or a ribozyme. An expression system consists of a gene, normally encoded by DNA, and the molecular machinery required to transcribe the DNA into mRNA and translate the mRNA into protein using the reagents provided. In the broadest sense this includes every living cell but the term is more normally used to refer to expression as a laboratory tool. An expression system is therefore often artificial in some manner. Expression systems are, however, a fundamentally natural process. Viruses are an excellent example where they replicate by using the host cell as an expression system for the viral proteins and genome.

In Nature

In addition to these biological tools, certain naturally observed configurations of DNA (genes, promoters, enhancers, repressors) and the associated machinery itself are referred to as an expression system. This term is normally used in the case where a gene or set of genes is switched on under well defined conditions. For example the simple repressor 'switch' expression system in Lambda phage and the lac operator system in bacteria. Several natural expression systems are directly used or modified and used for artificial expression systems such as the Tet-on and Tet-off expression system.

Gene Networks and Expression

Genes have sometimes been regarded as nodes in a network, with inputs being proteins such as transcription factors, and outputs being the level of gene expression. The node itself performs a function, and the operation of these functions have been interpreted as performing a kind of information processing within cell and determine cellular behavior.

Gene networks can also be constructed without formulating an explicit causal model. This is often the case when assembling networks from large expression data sets. Covariation and correlation of expression is computed across a large sample of cases and measurements (often transcriptome or proteome data). The source of variation can be either experimental or natural (observational). There are several ways to construct gene expression networks, but one common approach is to compute a matrix of all pair-wise correlations of expression across conditions, time points, or individuals and convert the matrix (after thresholding at some cut-off value) into a graphical representation in which nodes represent genes, transcripts, or proteins and edges connecting these nodes represent the strength of association.

Techniques and Tools

The following experimental techniques are used to measure gene expression and are listed in roughly chronological order, starting with the older, more established technologies. They are divided into two groups based on their degree of multiplexity.

- Low-to-mid-plex techniques:
  - o Reporter gene
  - o Northern blot
  - o Western blot
  - o Fluorescent in situ hybridization
  - o Reverse transcription PCR.
- Higher-plex techniques:
- o SAGE
- o DNA microarray
- o Tiling array
- o RNA-Seq.

Gene expression first involves transcription, in which DNA is used as a template to produce RNA. In the case of genes encoding proteins, that RNA produced from this process is messenger RNA (mRNA), which then needs to be translated by ribosomes to form a protein. As ribosomes are located outside the nucleus, mRNA produced needs to be exported.

Since the nucleus is the site of transcription, it also contains a variety of proteins which either directly mediate transcription or are involved in regulating the process. These proteins include helicases that unwind the double-stranded DNA molecule to facilitate access to it, RNA polymerases that synthesize the growing RNA molecule, topoisomerases that change the amount of supercoiling in DNA, helping it wind and unwind, as well as a large variety of transcription factors that regulate expression.

Processing of Pre-mRNA

Newly synthesized mRNA molecules are known as primary transcripts or pre-mRNA. They must undergo post-transcriptional modification in the nucleus before being exported to the cytoplasm; mRNA that appears in the nucleus without these modifications is degraded rather than used for protein translation. The three main modifications are 5' capping, 3' polyadenylation, and RNA splicing. While in the nucleus, pre-mRNA is associated with a variety of proteins in complexes known as heterogeneous ribonucleoprotein particles (hnRNPs). Addition of the 5' cap occurs co-transcriptionally and is the first step in post-transcriptional modification. The 3' poly-adenine tail is only added after transcription is complete.

RNA splicing, carried out by a complex called the spliceosome, is the process by which introns, or regions of DNA that do not code for protein, are removed from the pre-mRNA and the remaining exons connected to re-form a single continuous molecule. This process normally occurs after 5' capping and 3' polyadenylation but can begin before synthesis is complete in transcripts with many exons. Many pre-mRNAs, including those encoding antibodies, can be spliced in multiple ways to produce different mature mRNAs that encode different protein sequences. This process is known as alternative splicing, and allows production of a large variety of proteins from a limited amount of DNA.

Post-transcriptional Modification

Post-transcriptional modification is a process in cell biology by which, in eukaryotic cells, primary transcript RNA is converted into mature RNA. A notable example is the conversion of precursor messenger RNA into mature messenger RNA (mRNA), which includes splicing and occurs prior to protein synthesis. This process is vital for the correct translation of the genomes of eukaryotes as the human primary RNA transcript that is produced as a result of transcription contains both exons, which are coding sections of the primary RNA transcript and introns, which are the non coding sections of the primary RNA transcript.

Dynamics and Regulation

Nuclear Transport

The entry and exit of large molecules from the nucleus is tightly controlled by the nuclear pore complexes. Although small molecules can enter the nucleus without regulation, macromolecules such as RNA and proteins require association karyopherins called importins to enter the nucleus and exportins to exit. "Cargo" proteins that must be translocated from the cytoplasm to the nucleus contain short amino acid sequences known as nuclear localization signals which are bound by importins, while those transported from the nucleus to the cytoplasm carry nuclear export signals bound by exportins. The ability of importins and exportins to transport their cargo is regulated by GTPases, enzymes that hydrolyze the molecule guanosine triphosphate to release energy. The key GTPase in nuclear transport is Ran, which can bind either GTP or GDP (guanosine diphosphate) depending on whether it is located in the nucleus or the cytoplasm. Whereas importins depend on RanGTP to dissociate from their cargo, exportins require RanGTP in order to bind to their cargo.

Nuclear import depends on the importin binding its cargo in the cytoplasm and carrying it through the nuclear pore into the nucleus. Inside the nucleus, RanGTP acts to separate the cargo from the importin, allowing the importin to exit the nucleus and be reused. Nuclear export is similar, as the exportin binds the cargo inside the nucleus in a process facilitated by RanGTP, exits through the nuclear pore, and separates from its cargo in the cytoplasm.

Specialized export proteins exist for translocation of mature mRNA and tRNA to the cytoplasm after post-transcriptional modification is complete. This quality-control mechanism is important due to the these molecules' central role in protein translation; mis-expression of a protein due to incomplete excision of exons or mis-incorporation of amino acids could have negative consequences for the cell; thus incompletely modified RNA that reaches the cytoplasm is degraded rather than used in translation.

Assembly and Disassembly

During its lifetime a nucleus may be broken down, either in the process of cell division or as a consequence of apoptosis, a regulated form of cell death. During these events, the structural components of the nucleus—the envelope and lamina—are systematically degraded.

During the cell cycle the cell divides to form two cells. In order for this process to be possible, each of the new daughter cells must have a full set of genes, a process requiring replication of the chromosomes as well as segregation of the separate sets. This occurs by the replicated chromosomes, the sister chromatids, attaching to microtubules, which in turn are attached to different centrosomes. The sister chromatids can then be pulled to separate locations in the cell. In many cells the centrosome is located in the cytoplasm, outside the nucleus, the microtubules would be unable to attach to the chromatids in the presence of the nuclear envelope. Therefore the early stages in the cell cycle, beginning in prophase and until around prometaphase, the nuclear membrane is dismantled. Likewise, during the same period, the nuclear lamina is also disassembled, a process regulated by phosphorylation of the lamins. Towards the end of the cell cycle, the nuclear membrane is reformed, and around the same time, the nuclear lamina are reassembled by dephosphorylating the lamins.

However, in dinoflagellates the nuclear envelope remains intact, the centrosomes are located in the cytoplasm, and the microtubules come in contact with chromosomes, whose centromeric regions are incorporated into the nuclear envelope (the so-called closed mitosis with extranuclear spindle). In many other protists (e.g. ciliates, sporozoans) and fungi the centrosomes are intranuclear, and their nuclear envelope also does not disassemle during cell division.

Apoptosis is a controlled process in which the cell's structural components are destroyed, resulting in death of the cell. Changes associated with apoptosis directly affect the nucleus and its contents, for example in the condensation of chromatin and the disintegration of the nuclear envelope and lamina.

The destruction of the lamin networks is controlled by specialized apoptotic proteases called caspases, which cleave the lamin proteins and thus degrade the nucleus' structural integrity. Lamin cleavage is sometimes used as a laboratory indicator of caspase activity in assays for early apoptotic activity. Cells that express mutant caspase-resistant lamins are deficient in nuclear changes related to apoptosis, suggesting that lamins play a role in initiating the events that lead to apoptotic degradation of the nucleus. Inhibition of lamin assembly itself is an inducer of apoptosis.

The nuclear envelope acts as a barrier that prevents both DNA and RNA viruses from entering the nucleus. Some viruses require access to proteins inside the nucleus in order to replicate and/or assemble. DNA viruses, such as herpesvirus replicate and assemble in the cell nucleus, and exit by budding through the inner nuclear membrane. This process is accompanied by disassembly of the lamina on the nuclear face of the inner membrane.

Anucleated and Polynucleated Cells

Although most cells have a single nucleus, some eukaryotic cell types have no nucleus, and others have many nuclei. This can be a normal process, as in the maturation of mammalian red blood cells, or a result of faulty cell division.

Anucleated cells contain no nucleus and are therefore incapable of dividing to produce daughter cells. The best-known anucleated cell is the mammalian red blood cell, or erythrocyte, which also lacks other organelles such as mitochondria and serves primarily as a transport vessel to ferry oxygen from the lungs to the body's tissues. Erythrocytes mature through erythropoiesis in the bone marrow, where they lose their nuclei, organelles, and ribosomes. The nucleus is expelled during the process of differentiation from an erythroblast to a reticulocyte, which is the immediate precursor of the mature erythrocyte. The presence of mutagens may induce the release of some immature "micronucleated" erythrocytes into the bloodstream. Anucleated cells can also arise from flawed cell division in which one daughter lacks a nucleus and the other has two nuclei.

Polynucleated cells contain multiple nuclei. Most Acantharean species of protozoa and some fungi in mycorrhizae have naturally polynucleated cells. Other examples include the intestinal parasites in the genus *Giardia*, which have two nuclei per cell. In humans, skeletal muscle cells, called myocytes, become polynucleated during development; the resulting arrangement of nuclei near the periphery of the cells allows maximal intracellular space for myofibrils. Multinucleated cells can also be abnormal in humans; for example, cells arising from the fusion of monocytes and macrophages, known as giant multinucleated cells, sometimes accompany inflammation and are also implicated in tumour formation.

Evolution

As the major defining characteristic of the eukaryotic cell, the nucleus' evolutionary origin has been the subject of much speculation. Four major theories have been proposed to explain the existence of the nucleus, although none have yet earned widespread support.

The theory known as the "syntrophic model" proposes that a symbiotic relationship between the archaea and bacteria created the nucleus-containing eukaryotic cell. It is hypothesized that the symbiosis originated when ancient archaea, similar to modern methanogenic archaea, invaded and lived within bacteria similar to modern myxobacteria, eventually forming the early nucleus. This theory is analogous to the accepted theory for the origin of eukaryotic mitochondria and chloroplasts, which are thought to have developed from a similar endosymbiotic relationship between proto-eukaryotes and aerobic bacteria. The archaeal origin of the nucleus is supported by observations that archaea and eukarya have similar genes for certain proteins, including histones. Observations that myxobacteria are motile, can form multicellular complexes, and possess kinases and G proteins similar to eukarya, support a bacterial origin for the eukaryotic cell.

A second model proposes that proto-eukaryotic cells evolved from bacteria without an endosymbiotic stage. This model is based on the existence of modern planctomycetes bacteria that possess a nuclear structure with primitive pores and other compartmentalized membrane structures. A similar proposal states that a eukaryote-like cell, the chronocyte, evolved first and phagocytosed archaea and bacteria to generate the nucleus and the eukaryotic cell. The most controversial model, known as *viral eukaryogenesis*, posits that the membrane-bound nucleus, along with other eukaryotic features, originated from the infection of a prokaryote by a virus. The suggestion is based on similarities between eukaryotes and viruses such as linear DNA strands, mRNA capping, and tight binding to proteins (analogizing histones to viral envelopes). One version of the proposal suggests that the nucleus evolved in concert with phagocytosis to form an early cellular "predator". Another variant proposes that eukaryotes originated from early archaea infected by poxviruses, on the basis of observed similarity between the DNA polymerases in modern poxviruses and eukaryotes. It has been suggested that the unresolved question of the evolution of sex could be related to the viral eukaryogenesis hypothesis.

Finally, a very recent proposal suggests that traditional variants of the endosymbiont theory are insufficiently powerful to explain the origin of the eukaryotic nucleus. This model, termed the *exomembrane hypothesis*, suggests that the nucleus instead originated from a single ancestral cell that evolved a second exterior cell membrane; the interior membrane enclosing the original cell then became the nuclear membrane and evolved increasingly elaborate pore structures for passage of internally synthesized cellular components such as ribosomal subunits.

Chapter 4

Transfer Bioinorganic Chemistry

An electron transport chain (ETC) couples a reaction between an electron donor (such as NADH) and an electron acceptor (such as O_2) to the transfer of H^+ ions across a membrane, through a set of mediating biochemical reactions. These H^+ ions are used to produce adenosine triphosphate (ATP), the main energy intermediate in living organisms, as they move back across the membrane. Electron transport chains are used for extracting energy from sunlight (photosynthesis) and from redox reactions such as the oxidation of sugars (respiration).

In chloroplasts, light drives the conversion of water to oxygen and $NADP^+$ to NADPH and a transfer of H^+ ions. NADPH is used as an electron donor for carbon fixation. In mitochondria, it is the conversion of oxygen to water, NADH to NAD^+ and succinate to fumarate that drives the transfer of H^+ ions. While some bacteria have electron transport chains similar to those in chloroplasts or mitochondria, other bacteria use different electron donors and acceptors. Both the respiratory and photosynthetic electron transport chains are major sites of premature electron leakage to oxygen, thus being major sites of superoxide production and drivers of oxidative stress.

Background

The electron transport chain consists of a spatially separated series of redox reactions in which electrons are transferred from a donor molecule to an acceptor molecule. The underlying force driving these reactions is the Gibbs free energy of the reactants and products.

The Gibbs free energy is the energy available ("free") to do work. Any reaction that decreases the overall Gibbs free energy of a system will proceed spontaneously. The transfer of electrons proceeds from an electron donor to an acceptor.

ATP synthase, an enzyme highly conserved among all domains of life, is powered by a transmembrane proton electrochemical gradient, which is the result of a series of redox reactions. The function of the electron transport chain is to produce this gradient.

The transmembrane electrochemical potential gradient may enable transport of molecules across membranes. It may also enable mechanical work, such as rotating bacterial flagella, or to produce ATP, which provides energy to power other cellular reactions. A small amount of ATP is available from substrate-level phosphorylation, for example, in glycolysis. In most organisms the majority of ATP is generated in electron transport chains, while only some obtain ATP by fermentation.

Electron Transport Chains in Mitochondria

Most eukaryotic cells contain mitochondria, which produce ATP from products of the Krebs cycle, fatty acid oxidation, and amino acid oxidation. At the mitochondrial inner membrane, electrons from NADH and succinate pass through the electron transport chain to oxygen, which is reduced to water. This enzymatic series produces a proton gradient across the mitochondrial membrane, producing a thermodynamic state that has the potential to do work. A small percentage of electrons prematurely leak to oxygen, resulting in the formation of the toxic free-radical superoxide, a molecule thought to contribute to a number of diseases and aging.

Mitochondrial Redox Carriers

Stylized representation of the ETC. Energy obtained through the transfer of electrons (black arrows) down the ETC is used to pump protons (red arrows) from the mitochondrial matrix into the intermembrane space, creating an electrochemical proton gradient across the mitochondrial inner membrane (IMM) called $\Delta\Psi$. This electrochemical proton gradient allows ATP synthase (ATP-ase) to use the flow of H^+ through the enzyme back into the matrix to generate ATP from adenosine diphosphate (ADP) and inorganic phosphate. Complex I (NADH coenzyme Q reductase; labelled I) accepts electrons from the Krebs cycle electron carrier nicotinamide adenine dinucleotide (NADH), and passes them to coenzyme Q (ubiquinone; labelled UQ), which also receives electrons from complex II (succinate dehydrogenase; labelled II). UQ passes electrons to complex III (cytochrome bc_1 complex; labelled III), which passes them to cytochrome *c* (cyt *c*). Cyt *c* passes electrons to Complex IV (cytochrome *c* oxidase; labelled IV), which uses the electrons and hydrogen ions to reduce molecular oxygen to water.

Four membrane-bound complexes have been identified in mitochondria. Each is an extremely complex transmembrane structure that is embedded in the inner membrane. Three of them are proton pumps. The structures are electrically connected by lipid-soluble electron carriers and water-soluble electron carriers. The overall electron transport chain

NADH → *Complex I* → Q → *Complex III* → cytochrome *c* → *Complex IV* → O_2 → *Complex II*

Complex I

In *Complex I* (NADH dehydrogenase, also called NADH:ubiquinone oxidoreductase; EC 1.6.5.3) two electrons are removed from NADH and transferred to a lipid-soluble carrier, *ubiquinone* (Q). The reduced product, *ubiquinol* (QH_2) freely diffuses within the membrane, and Complex I translocates four protons (H^+) across the membrane, thus producing a proton gradient. Complex I is one of the main sites at which premature electron leakage to oxygen occurs, thus being one of main sites of production of harmful superoxide.

The pathway of electrons occurs as follows: NADH is oxidized to NAD^+, by reducing Flavin mononucleotide to $FMNH_2$ in one two-electron step. $FMNH_2$ is then oxidized in two one-electron steps, through a semiquinone intermediate. Each electron thus transfers from the $FMNH_2$ to an Fe-S cluster, from the Fe-S cluster to ubiquinone (Q). Transfer of the first electron results in the free-radical (semiquinone) form of Q, and transfer of the second electron reduces the semiquinone form to the ubiquinol form, QH_2. During this process, four protons are translocated from the mitochondrial matrix to the intermembrane space, creating a proton gradient that generates ATP through oxidative phosphorylation.

Complex II

In *Complex II* additional electrons are delivered into the quinone pool (Q) originating from succinate and transferred (via FAD) to Q. Complex II consists of four protein subunits: SDHA, SDHB, SDHC, and SDHD. Other electron donors (e.g., fatty acids and glycerol 3-phosphate) also direct electrons into Q (via FAD).

Complex III

In *Complex III* (cytochrome bc_1 complex; EC 1.10.2.2) two electrons are removed from QH_2 at the Q_O site and sequentially transferred to two molecules of cytochrome *c*, a water-soluble electron carrier located within the intermembrane space. The two other electrons sequentially pass across the protein to the Q_i site where the quinone part of ubiquinone is reduced to quinol.

A proton gradient is formed by two quinol (4H+4e-) oxidations at the Q_o site to form one quinol (2H+2e-) at the Q_i site. (in total six protons are translocated: two protons reduce quinone to quinol and four protons are released from two ubiquinol molecules). The bc1 complex does not 'pump' protons, but helps build the proton gradient by an asymmetric absorption/release of protons.

When electron transfer is reduced (by a high membrane potential or respiratory inhibitors such as antimycin A), Complex III may leak electrons to molecular oxygen, resulting in superoxide formation.

Complex IV

In *Complex IV* four electrons are removed from four molecules of cytochrome *c* and transferred to molecular oxygen (O_2), producing two molecules of water. At the same time, four protons are translocated across the membrane, contributing to the proton gradient. The activity of cytochrome c is inhibited by cyanide.

Coupling with Oxidative Phosphorylation

According to the chemiosmotic coupling hypothesis, proposed by Nobel Prize in Chemistry winner Peter D. Mitchell, the electron transport chain and oxidative phosphorylation are coupled by a proton gradient across the inner mitochondrial membrane. The efflux of protons from the mitochondrial matrix creates an electrochemical gradient (proton gradient). This gradient is used by the F_0F_1 ATP synthase complex to make ATP via oxidative phosphorylation. ATP synthase is sometimes described as *Complex V* of the electron transport chain. The F_0 component of ATP synthase acts as an ion channel that provides for a proton flux back into the mitochondrial matrix. This reflux releases free energy produced during the generation of the oxidized forms of the electron carriers (NAD^+ and Q). The free energy is used to drive ATP synthesis, catalysed by the F_1 component of the complex. Coupling with oxidative phosphorylation is a key step for ATP production. However, in certain cases, uncoupling the two processes may be biologically useful. The uncoupling protein, thermogenin—present in the inner mitochondrial membrane of brown adipose tissue—provides for an alternative flow of protons back to the inner mitochondrial matrix. This alternative flow results in thermogenesis rather than ATP production and generates heat.

Synthetic uncouplers (e.g., 2,4-dinitrophenol) also exist, and, at high doses, are lethal.

Summary

In the mitochondrial electron transport chain electrons move from an electron donor (NADH or QH_2) to a terminal electron acceptor (O_2) via a series of redox reactions. These reactions are coupled to the creation of a proton gradient across the mitochondrial inner membrane. There are three proton pumps: *I*, *III*, and *IV*. The resulting transmembrane proton gradient is used to make ATP via ATP synthase.

The reactions catalysed by *Complex I* and *Complex III* work roughly at equilibrium. This means that these reactions are readily reversible, by increasing the concentration of the products relative to the concentration of the reactants (for example, by increasing the proton gradient). ATP synthase is also readily reversible. Thus ATP can be used to build a proton gradient, which in turn can be used to make NADH. This process of *reverse electron transport* is important in many prokaryotic electron transport chains.

Electron Transport Chains in Bacteria

In eukaryotes, NADH is the most important electron donor. The associated electron transport chain is:

NADH → *Complex I* → Q → *Complex III* → cytochrome *c* → *Complex IV* → O_2 where *Complexes I, III* and *IV* are proton pumps, while Q and cytochrome *c* are mobile electron carriers. The electron acceptor is molecular oxygen. Individual bacteria use multiple electron transport chains, often simultaneously. Bacteria can use a number of different electron donors, a number of different dehydrogenases, a number of different oxidases and reductases, and a number of different electron acceptors. For example, *E. coli* (when growing aerobically using glucose as an energy source) uses two different NADH dehydrogenases and two different quinol oxidases, for a total of four different electron transport chains operating simultaneously.

A common feature of all electron transport chains is the presence of a proton pump to create a transmembrane proton gradient. Bacterial electron transport chains may contain as many as three proton pumps, like mitochondria, or they may contain only one or two. They always contain at least one proton pump.

Electron Donors

In the present day biosphere, the most common electron donors are organic molecules. Organisms that use organic molecules as an energy source are called *organotrophs*. Organotrophs (animals, fungi, protists) and *phototrophs* (plants and algae) constitute the vast majority of all familiar life forms.

Some prokaryotes can use inorganic matter as an energy source. Such organisms are called *lithotrophs* ("rock-eaters"). Inorganic electron donors include hydrogen, carbon monoxide, ammonia, nitrite, sulfur, sulfide, and ferrous iron.

Lithotrophs have been found growing in rock formations thousands of meters below the surface of Earth. Because of their volume of distribution, lithotrophs may actually outnumber organotrophs and phototrophs in our biosphere.

The use of inorganic electron donors as an energy source is of particular interest in the study of evolution. This type of metabolism must logically have preceded the use of organic molecules as an energy source.

Dehydrogenases

Bacteria can use a number of different electron donors. When organic matter is the energy source, the donor may be NADH or succinate, in which case electrons enter the electron transport chain via NADH dehydrogenase (similar to *Complex I* in mitochondria) or succinate dehydrogenase (similar to *Complex II*).

Other dehydrogenases may be used to process different energy sources: formate dehydrogenase, lactate dehydrogenase, glyceraldehyde-3-phosphate dehydrogenase, H_2 dehydrogenase (hydrogenase), etc. Some dehydrogenases are also proton pumps; others simply funnel electrons into the quinone pool. Most of dehydrogenases are synthesized only when needed. Depending on the environment in which they find themselves, bacteria select different enzymes from their DNA library and synthesize only those that are needed for growth. Enzymes that are synthesized only when needed are said to be 'inducible'.

Quinone Carriers

Quinones are mobile, lipid-soluble carriers that shuttle electrons (and protons) between large, relatively immobile macromolecular complexes embedded in the membrane. Bacteria use *ubiquinone* (the same quinone that mitochondria use) and related quinones such as *menaquinone*.

Proton Pumps

A *proton pump* is any process that creates a proton gradient across a membrane. Protons can be physically moved across a membrane;

this is seen in mitochondrial *Complexes I* and *IV*. The same effect can be produced by moving electrons in the opposite direction. The result is the disappearance of a proton from the cytoplasm and the appearance of a proton in the periplasm. Mitochondrial *Complex III* uses this second type of proton pump, which is mediated by a quinone (the Q cycle).

Some dehydrogenases are proton pumps; others are not. Most oxidases and reductases are proton pumps, but some are not. Cytochrome bc_1 is a proton pump found in many, but not all, bacteria (it is not found in *E. coli*). As the name implies, bacterial bc_1 is similar to mitochondrial bc_1 (*Complex III*). Proton pumps are the heart of the electron transport process. They produce the transmembrane electrochemical gradient that supplies energy to the cell.

Cytochrome Electron Carriers

Cytochromes are pigments that contain iron. They are found in two very different environments.

Some cytochromes are water-soluble carriers that shuttle electrons to and from large, immobile macromolecular structures imbedded in the membrane. The mobile cytochrome electron carrier in mitochondria is cytochrome *c*. Bacteria use a number of different mobile cytochrome electron carriers.

Other cytochromes are found within macromolecules such as *Complex III* and *Complex IV*. They also function as electron carriers, but in a very different, intramolecular, solid-state environment.

Electrons may enter an electron transport chain at the level of a mobile cytochrome or quinone carrier. For example, electrons from inorganic electron donors (nitrite, ferrous iron, etc.) enter the electron transport chain at the cytochrome level. When electrons enter at a redox level greater than NADH, the electron transport chain must operate in reverse to produce this necessary, higher-energy molecule.

Terminal Oxidases and Reductases

When bacteria grow in aerobic environments, the terminal electron acceptor (O_2) is reduced to water by an enzyme called an *oxidase*. When bacteria grow in anaerobic environments, the terminal electron acceptor is reduced by an enzyme called a *reductase*.

In mitochondria the terminal membrane complex (*Complex IV*) is cytochrome oxidase. Aerobic bacteria use a number of different terminal oxidases.

For example, *E. coli* does not have a cytochrome oxidase or a bc_1 complex. Under aerobic conditions, it uses two different terminal quinol oxidases (both proton pumps) to reduce oxygen to water.

Anaerobic bacteria, which do not use oxygen as a terminal electron acceptor, have terminal reductases individualized to their terminal acceptor. For example, *E. coli* can use fumarate reductase, nitrate reductase, nitrite reductase, DMSO reductase, or trimethylamine-N-oxide reductase, depending on the availability of these acceptors in the environment.

Most terminal oxidases and reductases are *inducible*. They are synthesized by the organism as needed, in response to specific environmental conditions.

Electron Acceptors

Just as there are a number of different electron donors (organic matter in organotrophs, inorganic matter in lithotrophs), there are a number of different electron acceptors, both organic and inorganic. If oxygen is available, it is invariably used as the terminal electron acceptor, because it generates the greatest Gibbs free energy change and produces the most energy.

In anaerobic environments, different electron acceptors are used, including nitrate, nitrite, ferric iron, sulfate, carbon dioxide, and small organic molecules such as fumarate. Since electron transport chains are redox processes, they can be described as the sum of two redox pairs. For example, the mitochondrial electron transport chain can be described as the sum of the NAD^+/NADH redox pair and the O_2/H_2O redox pair. NADH is the electron donor and O_2 is the electron acceptor.

Not every donor-acceptor combination is thermodynamically possible. The redox potential of the acceptor must be more positive than the redox potential of the donor. Furthermore, actual environmental conditions may be far different from *standard* conditions (1 molar concentrations, 1 atm partial pressures, pH = 7), which apply to *standard* redox potentials. For example, hydrogen-evolving bacteria grow at an ambient partial pressure of hydrogen gas of 10^{-4} atm. The associated redox reaction, which is thermodynamically favourable in nature, is thermodynamically impossible under "standard" conditions.

Bacterial electron transport pathways are, in general, inducible. Depending on their environment, bacteria can synthesize different transmembrane complexes and produce different electron transport chains in their cell membranes. Bacteria select their electron transport chains from a DNA library containing multiple possible dehydrogenases, terminal oxidases and terminal reductases. The situation is often summarized by saying that electron transport chains in bacteria are *branched*, *modular*, and *inducible*.

Photosynthetic Electron Transport Chains

In oxidative phosphorylation, electrons are transferred from a high-energy electron donor (e.g., NADH) to an electron acceptor (e.g., O_2) through an electron transport chain. In photophosphorylation, the energy of sunlight is used to *create* a high-energy electron donor and an electron acceptor. Electrons are then transferred from the donor to the acceptor through another electron transport chain.

Photosynthetic electron transport chains have many similarities to the oxidative chains. They use mobile, lipid-soluble carriers (quinones) and mobile, water-soluble carriers (cytochromes, etc.). They also contain a proton pump. It is remarkable that the proton pump in *all* photosynthetic chains resembles mitochondrial *Complex III.*

Photosynthetic electron transport chains are discussed in greater detail in the articles Photophosphorylation, Photosynthesis, Photosynthetic reaction centre and Light-dependent reaction. Electron transport chains are redox reactions that transfer electrons from an electron donor to an electron acceptor.

The transfer of electrons is coupled to the translocation of protons across a membrane, producing a proton gradient. The proton gradient is used to produce useful work.

The coupling of thermodynamically favourable to thermodynamically unfavourable biochemical reactions by biological macromolecules is an example of an emergent property – a property that could not have been predicted, even given full knowledge of the primitive geochemical systems from which these macromolecules evolved. It is an open question whether such emergent properties evolve only by chance, or whether they *necessarily* evolve in any large biogeochemical system, given the underlying laws of physics.

Atom and Group Transfer Chemistry

The atom is a basic unit of matter that consists of a dense, central nucleus surrounded by a cloud of negatively charged electrons. The atomic nucleus contains a mix of positively charged protons and electrically neutral neutrons (except in the case of hydrogen-1, which is the only stable nuclide with no neutrons). The electrons of an atom are bound to the nucleus by the electromagnetic force.

Likewise, a group of atoms can remain bound to each other, forming a molecule. An atom containing an equal number of protons and electrons is electrically neutral, otherwise it has a positive or negative charge and is an ion. An atom is classified according to the number of protons and neutrons in its nucleus: the number of protons determines the chemical element, and the number of neutrons determines the isotope of the element.

The concept of an atom as an indivisible component of matter was first proposed by early Indian and Greek philosophers. In the 17th and 18th centuries, chemists provided a physical basis for this idea by showing that certain substances could not be further broken down by chemical methods. During the late 19th and early 20th centuries, physicists discovered subatomic components and structure inside the atom, thereby demonstrating that the 'atom' was divisible. The principles of quantum mechanics were used to successfully model the atom.

Atoms are minuscule objects with proportionately tiny masses. Atoms can only be observed individually using special instruments such as the scanning tunnelling microscope. Over 99.9% of an atom's mass is concentrated in the nucleus, with protons and neutrons having roughly equal mass. Each element has at least one isotope with unstable nuclei that can undergo radioactive decay. This can result in a transmutation that changes the number of protons or neutrons in a nucleus. Electrons that are bound to atoms possess a set of stable energy levels, or orbitals, and can undergo transitions between them by absorbing or emitting photons that match the energy differences between the levels. The electrons determine the chemical properties of an element, and strongly influence an atom's magnetic properties.

History

Atomism

The concept that matter is composed of discrete units and cannot be divided into arbitrarily tiny quantities has been around for millennia, but these ideas were founded in abstract, philosophical reasoning rather than experimentation and empirical observation. The nature of atoms in philosophy varied considerably over time and between cultures and schools, and often had spiritual elements. Nevertheless, the basic idea of the atom was adopted by scientists thousands of years later because it elegantly explained new discoveries in the field of chemistry. The earliest references to the concept of atoms date back to ancient India in the 6th century BCE, appearing first in Jainism. The Nyaya and Vaisheshika schools developed elaborate theories of how atoms combined into more complex objects. In the West, the references to atoms emerged a century later from Leucippus, whose student, Democritus, systematized his views. In approximately 450 BCE, Democritus coined the term *átomos*, which means "uncuttable" or "the smallest indivisible particle of matter". Although the Indian and Greek concepts of the atom were based purely on philosophy, modern science has retained the name coined by Democritus.

Corpuscularianism is the postulate, expounded in the 13th-century by the alchemist Pseudo-Geber (Geber), sometimes identified with Paul of Taranto, that all physical bodies possess an inner and outer layer of minute particles or corpuscles. Corpuscularianism is similar to the theory atomism, except that where atoms were supposed to be indivisible, corpuscles could in principle be divided. In this manner, for example, it was theorized that mercury could penetrate into metals and modify their inner structure. Corpuscularianism stayed a dominant theory over the next several hundred years.

In 1661, natural philosopher Robert Boyle published *The Sceptical Chymist* in which he argued that matter was composed of various combinations of different "corpuscules" or atoms, rather than the classical elements of air, earth, fire and water. During the 1670s was used by Isaac Newton in his development of the corpuscular theory of light.

Origin of Scientific Theory

Further progress in the understanding of atoms did not occur until the science of chemistry began to develop. In 1789, French nobleman and scientific researcher Antoine Lavoisier discovered the law of conservation of mass and defined an element as a basic substance that could not be further broken down by the methods of chemistry.

In 1803, English instructor and natural philosopher John Dalton used the concept of atoms to explain why elements always react in ratios of small whole numbers (the law of multiple proportions) and why certain gases dissolve better in water than others. He proposed that each element consists of atoms of a single, unique type, and that these atoms can join together to form chemical compounds. Dalton is considered the originator of modern atomic theory.

Dalton's atomic hypothesis did not specify the size of atoms. Common sense indicated they must be very small, but nobody knew how small. Therefore it was a major landmark when in 1865 Johann Josef Loschmidt measured the size of the molecules that make up air.

An additional line of reasoning in support of particle theory (and by extension atomic theory) began in 1827 when botanist Robert Brown used a microscope to look at dust grains floating in water and discovered that they moved about erratically—a phenomenon that became known as "Brownian motion". J. Desaulx suggested in 1877 that the phenomenon was caused by the thermal motion of water molecules, and in 1905 Albert Einstein produced the first mathematical analysis of the motion. French physicist Jean Perrin used Einstein's work to experimentally determine the mass and dimensions of atoms, thereby conclusively verifying Dalton's atomic theory.

In 1869, building upon earlier discoveries by such scientists as Lavoisier, Dmitri Mendeleev published the first functional periodic table. The table itself is a visual representation of the periodic law, which states that certain chemical properties of elements repeat *periodically* when arranged by atomic number.

Subcomponents and Quantum Theory

The physicist J. J. Thomson, through his work on cathode rays in 1897, discovered the electron, and concluded that they were a component of every atom. Thus he overturned the belief that atoms are the indivisible, ultimate particles of matter. Thomson postulated that the low mass, negatively charged electrons were distributed throughout the atom, possibly rotating in rings, with their charge balanced by the presence of a uniform sea of positive charge. This later became known as the plum pudding model.

In 1909, Hans Geiger and Ernest Marsden, under the direction of physicist Ernest Rutherford, bombarded a sheet of gold foil with alpha rays—by then known to be positively charged helium atoms—and discovered that a small percentage of these particles were deflected through much larger angles than was predicted using Thomson's proposal. Rutherford interpreted the gold foil experiment as suggesting that the positive charge of a heavy gold atom and most of its mass was concentrated in a nucleus at the centre of the atom—the Rutherford model.

While experimenting with the products of radioactive decay, in 1913 radiochemist Frederick Soddy discovered that there appeared to be more than one type of atom at each position on the periodic table. The term isotope was coined by Margaret Todd as a suitable name for different atoms that belong to the same element. J.J. Thomson created a technique for separating atom types through his work on ionized gases, which subsequently led to the discovery of stable isotopes.

Meanwhile, in 1913, physicist Niels Bohr suggested that the electrons were confined into clearly defined, quantized orbits, and could jump between these, but could not freely spiral inward or outward in intermediate states.

An electron must absorb or emit specific amounts of energy to transition between these fixed orbits. When the light from a heated material was passed through a prism, it produced a multi-coloured spectrum. The appearance of fixed lines in this spectrum was successfully explained by these orbital transitions.

Chemical bonds between atoms were now explained, by Gilbert Newton Lewis in 1916, as the interactions between their constituent electrons. As the chemical properties of the elements were known to largely repeat themselves according to the periodic law, in 1919 the American chemist Irving Langmuir suggested that this could be explained if the electrons in an atom were connected or clustered in some manner. Groups of electrons were thought to occupy a set of electron shells about the nucleus.

The Stern–Gerlach experiment of 1922 provided further evidence of the quantum nature of the atom. When a beam of silver atoms was passed through a specially shaped magnetic field, the beam was split based on the direction of an atom's angular momentum, or spin. As this direction is random, the beam could be expected to spread into a line. Instead, the beam was split into two parts, depending on whether the atomic spin was oriented up or down.

In 1924, Louis de Broglie proposed that all particles behave to an extent like waves. In 1926, Erwin Schrodinger used this idea to develop a mathematical model of the atom that described the electrons as three-dimensional waveforms rather than point particles. A consequence of using waveforms to describe particles is that it is mathematically impossible to obtain precise values for both the position and momentum of a particle at the same time; this became known as the uncertainty principle, formulated by Werner Heisenberg in 1926.

In this concept, for a given accuracy in measuring a position one could only obtain a range of probable values for momentum, and vice versa. This model was able to explain observations of atomic behaviour that previous models could not, such as certain structural and spectral patterns of atoms larger than hydrogen. Thus, the planetary model of the atom was discarded in favour of one that described atomic orbital zones around the nucleus where a given electron is most likely to be observed.

The development of the mass spectrometer allowed the exact mass of atoms to be measured. The device uses a magnet to bend the trajectory of a beam of ions, and the amount of deflection is determined by the ratio of an atom's mass to its charge. The chemist Francis William Aston used this instrument to show that isotopes had different masses.

The atomic mass of these isotopes varied by integer amounts, called the whole number rule. The explanation for these different isotopes awaited the discovery of the neutron, a neutral-charged particle with a mass similar to the proton, by the physicist James Chadwick in 1932. Isotopes were then explained as elements with the same number of protons, but different numbers of neutrons within the nucleus.

Fission, High Energy Physics and Condensed Matter

In 1938, the German chemist Otto Hahn, a student of Rutherford, directed neutrons onto uranium atoms expecting to get transuranium elements. Instead, his chemical experiments showed barium as a product. A year later, Lise Meitner and her nephew Otto Frisch verified that Hahn's result were the first experimental *nuclear fission*. In 1944, Hahn received the Nobel prize in chemistry. Despite Hahn's efforts, the contributions of Meitner and Frisch were not recognized.

In the 1950s, the development of improved particle accelerators and particle detectors allowed scientists to study the impacts of atoms moving at high energies. Neutrons and protons were found to be hadrons, or composites of smaller particles called quarks. Standard models of nuclear physics were developed that successfully explained the properties of the nucleus in terms of these sub-atomic particles and the forces that govern their interactions.

Components

Subatomic Particles

Though the word *atom* originally denoted a particle that cannot be cut into smaller particles, in modern scientific usage the atom is composed of various subatomic particles. The constituent particles of an atom are the electron, the proton and the neutron. However, the hydrogen-1 atom has no neutrons and a positive hydrogen ion has no electrons.

The electron is by far the least massive of these particles at 9.11×10 kg, with a negative electrical charge and a size that is too small to be measured using available techniques. Protons have a positive charge and a mass 1,836 times that of the electron, at 1.6726×10^{-27} kg, although this can be reduced by changes to the energy binding the proton into an atom. Neutrons have no electrical charge and have a free mass of 1,839 times the mass of electrons, or 1.6929×10^{-27} kg. Neutrons and protons have comparable dimensions—on the order of 2.5×10^{-15} m—although the 'surface' of these particles is not sharply defined.

In the Standard Model of physics, both protons and neutrons are composed of elementary particles called quarks. The quark belongs to the fermion group of particles, and is one of the two basic constituents of matter—the other being the lepton, of which the electron is an example. There are six types of quarks, each having a fractional electric charge of either +2/3 or "1/3. Protons are composed of two up quarks and one down quark, while a neutron consists of one up quark and two down quarks. This distinction accounts for the difference in mass and charge between the two particles. The quarks are held together by the strong nuclear force, which is mediated by gluons. The gluon is a member of the family of gauge bosons, which are elementary particles that mediate physical forces.

Nucleus

All the bound protons and neutrons in an atom make up a tiny atomic nucleus, and are collectively called nucleons. The radius of a nucleus is approximately equal to fm, where *A* is the total number of nucleons. This is much smaller than the radius of the atom, which is on the order of 10^5 fm. The nucleons are bound together by a short-ranged attractive potential called the residual strong force. At distances smaller than 2.5 fm this force is much more powerful than the electrostatic force that causes positively charged protons to repel each other.

Atoms of the same element have the same number of protons, called the atomic number. Within a single element, the number of neutrons may vary, determining the isotope of that element. The total number of protons and neutrons determine the nuclide. The number of neutrons relative to the protons determines the stability of the nucleus, with certain isotopes undergoing radioactive decay.

The neutron and the proton are different types of fermions. The Pauli exclusion principle is a quantum mechanical effect that prohibits *identical* fermions, such as multiple protons, from occupying the same quantum physical state at the same time. Thus every proton in the nucleus must occupy a different state, with its own energy level, and the same rule applies to all of the neutrons. This prohibition does not apply to a proton and neutron occupying the same quantum state.

For atoms with low atomic numbers, a nucleus that has a different number of protons than neutrons can potentially drop to a lower energy state through a radioactive decay that causes the number of protons and neutrons to more closely match. As a result, atoms with roughly matching numbers of protons and neutrons are more stable against decay. However, with increasing atomic number, the mutual repulsion of the protons requires an increasing proportion of neutrons to maintain the stability of the nucleus, which modifies this trend. Thus, there are no stable nuclei with equal proton and neutron numbers above atomic number Z = 20 (calcium); and as Z increases toward the heaviest nuclei, the ratio of neutrons per proton required for stability increases to about 1.5.

The number of protons and neutrons in the atomic nucleus can be modified, although this can require very high energies because of the strong force. Nuclear fusion occurs when multiple atomic particles join to form a heavier nucleus, such as through the energetic collision of two nuclei. For example, at the core of the Sun protons require energies of 3–10 keV to overcome their mutual repulsion—the coulomb barrier—and fuse together into a single nucleus. Nuclear fission is the opposite process, causing a nucleus to split into two smaller nuclei—usually through radioactive decay. The nucleus can also be modified through bombardment by high energy subatomic particles or photons. If this modifies the number of protons in a nucleus, the atom changes to a different chemical element.

If the mass of the nucleus following a fusion reaction is less than the sum of the masses of the separate particles, then the difference between these two values can be emitted as a type of usable energy (such as a gamma ray, or the kinetic energy of a beta particle), as described by Albert Einstein's mass–energy equivalence formula, $E = mc^2$, where m is the mass loss and c is the speed of light. This deficit is part of the binding energy of the new nucleus, and it is the non-recoverable loss of the energy that causes the fused particles to remain together in a state that requires this energy to separate.

The fusion of two nuclei that create larger nuclei with lower atomic numbers than iron and nickel—a total nucleon number of about 60—is usually an exothermic process that releases more energy than is required to bring them together. It is this energy-releasing process that makes nuclear fusion in stars a self-sustaining reaction. For heavier nuclei, the binding energy per nucleon in the nucleus begins to decrease. That means fusion processes producing nuclei that have atomic numbers higher than about 26, and atomic masses higher than about 60, is an endothermic process. These more massive nuclei can not undergo an energy-producing fusion reaction that can sustain the hydrostatic equilibrium of a star.

Electron Cloud

The electrons in an atom are attracted to the protons in the nucleus by the electromagnetic force. This force binds the electrons inside an electrostatic potential well surrounding the smaller nucleus, which means that an external source of energy is needed for the electron to escape. The closer an electron is to the nucleus, the greater the attractive force. Hence electrons bound near the centre of the potential well require more energy to escape than those at greater separations.

Electrons, like other particles, have properties of both a particle and a wave. The electron cloud is a region inside the potential well where each electron forms a type of three-dimensional standing wave—a wave form that does not move relative to the nucleus.

This behaviour is defined by an atomic orbital, a mathematical function that characterises the probability that an electron appears to be at a particular location when its position is measured. Only a discrete (or quantized) set of these orbitals exist around the nucleus, as other possible wave patterns rapidly decay into a more stable form. Orbitals can have one or more ring or node structures, and they differ from each other in size, shape and orientation. Each atomic orbital corresponds to a particular energy level of the electron. The electron can change its state to a higher energy level by absorbing a photon with sufficient energy to boost it into the new quantum state. Likewise, through spontaneous emission, an electron in a higher energy state can drop to a lower energy state while radiating the excess energy as a photon. These characteristic energy values, defined by the differences in the energies of the quantum states, are responsible for atomic spectral lines. The amount of energy needed to remove or add an electron—the electron binding energy—is far less than the binding energy of nucleons.

For example, it requires only 13.6 eV to strip a ground-state electron from a hydrogen atom, compared to 2.23 *million* eV for splitting a deuterium nucleus. Atoms are electrically neutral if they have an equal number of protons and electrons. Atoms that have either a deficit or a surplus of electrons are called ions. Electrons that are farthest from the nucleus may be transferred to other nearby atoms or shared between atoms. By this mechanism, atoms are able to bond into molecules and other types of chemical compounds like ionic and covalent network crystals.

Properties

Nuclear Properties

By definition, any two atoms with an identical number of *protons* in their nuclei belong to the same chemical element. Atoms with equal numbers of protons but a different number of *neutrons* are different isotopes of the same element. For example, all hydrogen atoms admit exactly one proton, but isotopes exist with no neutrons hydrogen-1, one neutron (deuterium), two neutrons (tritium) and more than two neutrons. The hydrogen-1 is by far the most common form, and is sometimes called protium. The known elements form a set of atomic numbers from hydrogen with a single proton up to the 118-proton element ununoctium. All known isotopes of elements with atomic numbers greater than 82 are radioactive.

About 339 nuclides occur naturally on Earth, of which 227 (about 67%) have not been observed to decay, and are referred to as "stable isotopes". However, only 90 of these nuclides are stable to all decay, even in theory. About 30 more (bringing the total to 257) have been observed to decay, but have half lives too long to be estimated. These are also often classified as "stable." An additional 31 radioactive nuclides have half lives longer than 80 million years, and are thus long-lived enough to be present from the birth of the solar system. This collection of 288 nuclides are known as primordial nuclides. Finally, an additional 51 short-lived nuclides are known to occur naturally, as daughter products of primordial nuclide decay (such as radium from uranium), or else as products of natural energetic processes on Earth, such as cosmic ray bombardment (for example, carbon-14).

For 80 of the chemical elements, at least one stable isotope exists. Elements 43, 61, and all elements numbered 83 or higher have no stable isotopes. As a rule, there is, for each element, only a handful of stable isotopes, the average being 3.1 stable isotopes per element among those that have stable isotopes. Twenty-seven elements have only a single stable isotope, while the largest number of stable isotopes observed for any element is ten, for the element tin.

Stability of isotopes is affected by the ratio of protons to neutrons, and also by the presence of certain "magic numbers" of neutrons or protons that represent closed and filled quantum shells. These quantum shells correspond to a set of energy levels within the shell model of the nucleus; filled shells, such as the filled shell of 50 protons for tin, confers unusual stability on the nuclide. Of the 256 known stable nuclides, only four have both an odd number of protons *and* odd number of neutrons: hydrogen-2 (deuterium), lithium-6, boron-10 and nitrogen-14. Also, only four naturally occurring, radioactive odd-odd nuclides have a half-life over a billion years: potassium-40, vanadium-50, lanthanum-138 and tantalum-180m. Most odd-odd nuclei are highly unstable with respect to beta decay, because the decay products are even-even, and are therefore more strongly bound, due to nuclear pairing effects.

Mass

The large majority of an atom's mass comes from the protons and neutrons, the total number of these particles in an atom is called the mass number. The mass of an atom at rest is often expressed using the unified atomic mass unit (u), which is also called a Dalton (Da). This unit is defined as a twelfth of the mass of a free neutral atom of carbon-12, which is approximately 1.66×10^{-27} kg. Hydrogen-1, the lightest isotope of hydrogen and the atom with the lowest mass, has an atomic weight of 1.007825 u. An atom has a mass approximately equal to the mass number times the atomic mass unit. The heaviest stable atom is lead-208, with a mass of 207.9766521 u.

As even the most massive atoms are far too light to work with directly, chemists instead use the unit of moles. The mole is defined such that one mole of any element always has the same number of atoms (about 6.022×10^{23}). This number was chosen so that if an element has an atomic mass of 1 u, a mole of atoms of that element has a mass close to 0.001 kg, or 1 gram. Because of the definition of the unified atomic mass unit, carbon has an atomic mass of *exactly* 12 u, and so a mole of carbon atoms weighs exactly 0.012 kg.

Shape and Size

Atoms lack a well-defined outer boundary, so their dimensions are usually described in terms of an atomic radius. This is a measure of the distance out to which the electron cloud extends from the nucleus. However, this assumes the atom to exhibit a spherical shape, which is only obeyed for atoms in vacuum or free space. Atomic radii may be derived from the distances between two nuclei when the two atoms are joined in a chemical bond. The radius varies with the location of an atom on the atomic chart, the type of chemical bond, the number of neighbouring atoms (coordination number) and a quantum mechanical property known as spin. On the periodic table of the elements, atom size tends to increase when moving down columns, but decrease when moving across rows. Consequently, the smallest atom is helium with a radius of 32 pm, while one of the largest is caesium at 225 pm.

When subjected to external fields, like an electrical field, the shape of an atom may deviate from that a sphere. The deformation depends on the field magnitude and the orbital type of outer shell electrons, as shown by group-theoretical considerations. Aspherical deviations might be elicited for instance in crystals, where large crystal-electrical fields may occur at low-symmetry lattice sites. Significant ellipsoidal deformations have recently been shown to occur for sulfur ions in pyrite-type compounds. Atomic dimensions are thousands of times smaller than the wavelengths of light (400–700 nm) so they can not be viewed using an optical microscope.

However, individual atoms can be observed using a scanning tunnelling microscope. To visualize the minuteness of the atom, consider that a typical human hair is about 1 million carbon atoms in width. A single drop of water contains about 2 sextillion (2×10^{21}) atoms of oxygen, and twice the number of hydrogen atoms. A single carat diamond with a mass of 2×10^{-4} kg contains about 10 sextillion (10^{22}) atoms of carbon. If an apple were magnified to the size of the Earth, then the atoms in the apple would be approximately the size of the original apple.

Radioactive Decay

Every element has one or more isotopes that have unstable nuclei that are subject to radioactive decay, causing the nucleus to emit particles or electromagnetic radiation. Radioactivity can occur when the radius of a nucleus is large compared with the radius of the strong force, which only acts over distances on the order of 1 fm.

The most common forms of radioactive decay are:

- Alpha decay is caused when the nucleus emits an alpha particle, which is a helium nucleus consisting of two protons and two neutrons. The result of the emission is a new element with a lower atomic number.
- Beta decay is regulated by the weak force, and results from a transformation of a neutron into a proton, or a proton into a neutron. The first is accompanied by the emission of an electron and an antineutrino, while the second causes the emission of a positron and a neutrino. The electron or positron emissions are called beta particles. Beta decay either increases or decreases the atomic number of the nucleus by one.
- Gamma decay results from a change in the energy level of the nucleus to a lower state, resulting in the emission of electromagnetic radiation. This can occur following the emission of an alpha or a beta particle from radioactive decay.

Other more rare types of radioactive decay include ejection of neutrons or protons or clusters of nucleons from a nucleus, or more than one beta particle, or result (through internal conversion) in production of high-speed electrons that are not beta rays, and high-energy photons that are not gamma rays.

Each radioactive isotope has a characteristic decay time period—the half-life—that is determined by the amount of time needed for half of a sample to decay. This is an exponential decay process that steadily decreases the proportion of the remaining isotope by 50% every half life. Hence after two half-lives have passed only 25% of the isotope is present, and so forth.

Magnetic Moment

Elementary particles possess an intrinsic quantum mechanical property known as spin. This is analogous to the angular momentum of an object that is spinning around its centre of mass, although strictly speaking these particles are believed to be point-like and cannot be said to be rotating. Spin is measured in units of the reduced Planck constant (‘), with electrons, protons and neutrons all having spin ½ ’, or “spin-½”. In an atom, electrons in motion around the nucleus possess orbital angular momentum in addition to their spin, while the nucleus itself possesses angular momentum due to its nuclear spin.

The magnetic field produced by an atom—its magnetic moment—is determined by these various forms of angular momentum, just as a rotating charged object classically produces a magnetic field. However, the most dominant contribution comes from spin. Due to the nature of electrons to obey the Pauli exclusion principle, in which no two electrons may be found in the same quantum state, bound electrons pair up with each other, with one member of each pair in a spin up state and the other in the opposite, spin down state. Thus these spins cancel each other out, reducing the total magnetic dipole moment to zero in some atoms with even number of electrons.

In ferromagnetic elements such as iron, an odd number of electrons leads to an unpaired electron and a net overall magnetic moment. The orbitals of neighbouring atoms overlap and a lower energy state is achieved when the spins of unpaired electrons are aligned with each other, a process known as an exchange interaction. When the magnetic moments of ferromagnetic atoms are lined up, the material can produce a measurable macroscopic field. Paramagnetic materials have atoms with magnetic moments that line up in random directions when no magnetic field is present, but the magnetic moments of the individual atoms line up in the presence of a field.

The nucleus of an atom can also have a net spin. Normally these nuclei are aligned in random directions because of thermal equilibrium. However, for certain elements (such as xenon-129) it is possible to polarize a significant proportion of the nuclear spin states so that they are aligned in the same direction—a condition called hyperpolarization. This has important applications in magnetic resonance imaging.

Energy Levels

When an electron is bound to an atom, it has a potential energy that is inversely proportional to its distance from the nucleus. This is measured by the amount of energy needed to unbind the electron from the atom, and is usually given in units of electronvolts (eV). In the quantum mechanical model, a bound electron can only occupy a set of states centered on the nucleus, and each state corresponds to a specific energy level. The lowest energy state of a bound electron is called the ground state, while an electron at a higher energy level is in an excited state.

For an electron to transition between two different states, it must absorb or emit a photon at an energy matching the difference in the potential energy of those levels. The energy of an emitted photon is proportional to its frequency, so these specific energy levels appear as distinct bands in the electromagnetic spectrum. Each element has a characteristic spectrum that can depend on the nuclear charge, subshells filled by electrons, the electromagnetic interactions between the electrons and other factors.

When a continuous spectrum of energy is passed through a gas or plasma, some of the photons are absorbed by atoms, causing electrons to change their energy level. Those excited electrons that remain bound to their atom spontaneously emit this energy as a photon, travelling in a random direction, and so drop back to lower energy levels. Thus the atoms behave like a filter that forms a series of dark absorption bands in the energy output. (An observer viewing the atoms from a view that doesn't include the continuous spectrum in the background, instead sees a series of emission lines from the photons emitted by the atoms.) Spectroscopic measurements of the strength and width of spectral lines allow the composition and physical properties of a substance to be determined.

Close examination of the spectral lines reveals that some display a fine structure splitting. This occurs because of spin-orbit coupling, which is an interaction between the spin and motion of the outermost electron. When an atom is in an external magnetic field, spectral lines become split into three or more components; a phenomenon called the Zeeman effect. This is caused by the interaction of the magnetic field with the magnetic moment of the atom and its electrons. Some atoms can have multiple electron configurations with the same energy level, which thus appear as a single spectral line. The interaction of the magnetic field with the atom shifts these electron configurations to slightly different energy levels, resulting in multiple spectral lines.

The presence of an external electric field can cause a comparable splitting and shifting of spectral lines by modifying the electron energy levels, a phenomenon called the Stark effect.

If a bound electron is in an excited state, an interacting photon with the proper energy can cause stimulated emission of a photon with a matching energy level. For this to occur, the electron must drop to a lower energy state that has an energy difference matching the energy of the interacting photon. The emitted photon and the interacting photon then move off in parallel and with matching phases. That is, the wave patterns of the two photons are synchronized. This physical property is used to make lasers, which can emit a coherent beam of light energy in a narrow frequency band.

Valence and Bonding Behaviour

The outermost electron shell of an atom in its uncombined state is known as the valence shell, and the electrons in that shell are called valence electrons. The number of valence electrons determines the bonding behaviour with other atoms. Atoms tend to chemically react with each other in a manner that fills (or empties) their outer valence shells. For example, a transfer of a single electron between atoms is a useful approximation for bonds that form between atoms with one-electron more than a filled shell, and others that are one-electron short of a full shell, such as occurs in the compound sodium chloride and other chemical ionic salts. However, many elements display multiple valences, or tendencies to share differing numbers of electrons in different compounds. Thus, chemical bonding between these elements takes many forms of electron-sharing that are more than simple electron transfers. Examples include the element carbon and the organic compounds.

The chemical elements are often displayed in a periodic table that is laid out to display recurring chemical properties, and elements with the same number of valence electrons form a group that is aligned in the same column of the table. (The horizontal rows correspond to the filling of a quantum shell of electrons.) The elements at the far right of the table have their outer shell completely filled with electrons, which results in chemically inert elements known as the noble gases.

States

Quantities of atoms are found in different states of matter that depend on the physical conditions, such as temperature and pressure. By varying the conditions, materials can transition between solids, liquids, gases and plasmas. Within a state, a material can also exist in different phases. An example of this is solid carbon, which can exist as graphite or diamond. At temperatures close to absolute zero, atoms can form a Bose–Einstein condensate, at which point quantum mechanical effects, which are normally only observed at the atomic scale, become apparent on a macroscopic scale. This super-cooled collection of atoms then behaves as a single super atom, which may allow fundamental checks of quantum mechanical behaviour.

Identification

The scanning tunnelling microscope is a device for viewing surfaces at the atomic level. It uses the quantum tunnelling phenomenon, which allows particles to pass through a barrier that would normally be insurmountable. Electrons tunnel through the vacuum between two planar metal electrodes, on each of which is an adsorbed atom, providing a tunnelling-current density that can be measured.

Scanning one atom (taken as the tip) as it moves past the other (the sample) permits plotting of tip displacement versus lateral separation for a constant current. The calculation shows the extent to which scanning-tunnelling-microscope images of an individual atom are visible. It confirms that for low bias, the microscope images the space-averaged dimensions of the electron orbitals across closely packed energy levels—the Fermi level local density of states.

An atom can be ionized by removing one of its electrons. The electric charge causes the trajectory of an atom to bend when it passes through a magnetic field. The radius by which the trajectory of a moving ion is turned by the magnetic field is determined by the mass of the atom. The mass spectrometer uses this principle to measure the mass-to-charge ratio of ions. If a sample contains multiple isotopes, the mass spectrometer can determine the proportion of each isotope in the sample by measuring the intensity of the different beams of ions. Techniques to vaporize atoms include inductively coupled plasma atomic emission spectroscopy and inductively coupled plasma mass spectrometry, both of which use a plasma to vaporize samples for analysis.

A more area-selective method is electron energy loss spectroscopy, which measures the energy loss of an electron beam within a transmission electron microscope when it interacts with a portion of a sample. The atom-probe tomograph has sub-nanometer resolution in 3-D and can chemically identify individual atoms using time-of-flight mass spectrometry.

Spectra of excited states can be used to analyze the atomic composition of distant stars. Specific light wavelengths contained in the observed light from stars can be separated out and related to the quantized transitions in free gas atoms. These colours can be replicated using a gas-discharge lamp containing the same element. Helium was discovered in this way in the spectrum of the Sun 23 years before it was found on Earth.

Origin and Current State

Atoms form about 4% of the total energy density of the observable universe, with an average density of about 0.25 atoms/m^3. Within a galaxy such as the Milky Way, atoms have a much higher concentration, with the density of matter in the interstellar medium (ISM) ranging from 10^5 to 10^9 atoms/m^3. The Sun is believed to be inside the Local Bubble, a region of highly ionized gas, so the density in the solar neighborhood is only about 10^3 atoms/m^3. Stars form from dense clouds in the ISM, and the evolutionary processes of stars result in the steady enrichment of the ISM with elements more massive than hydrogen and helium. Up to 95% of the Milky Way's atoms are concentrated inside stars and the total mass of atoms forms about 10% of the mass of the galaxy. (The remainder of the mass is an unknown dark matter.)

Nucleosynthesis

Stable protons and electrons appeared one second after the Big Bang. During the following three minutes, Big Bang nucleosynthesis produced most of the helium, lithium, and deuterium in the universe, and perhaps some of the beryllium and boron. The first atoms (complete with bound electrons) were theoretically created 380,000 years after the Big Bang—an epoch called recombination, when the expanding universe cooled enough to allow electrons to become attached to nuclei. Since then, atomic nuclei have been combined in stars through the process of nuclear fusion to produce elements up to iron.

Isotopes such as lithium-6 are generated in space through cosmic ray spallation. This occurs when a high-energy proton strikes an atomic nucleus, causing large numbers of nucleons to be ejected. Elements heavier than iron were produced in supernovae through the r-process and in AGB stars through the s-process, both of which involve the capture of neutrons by atomic nuclei. Elements such as lead formed largely through the radioactive decay of heavier elements.

Chapter 5

Chemistry of Bioorganic Products

Organic Food

Organic foods are made in a way that complies with organic standards set by national governments and international organizations. In the United States, organic production is a system that is managed in accordance with the Organic Foods Production Act (OFPA) of 1990 and regulations in Title 7, Part 205 of the Code of Federal Regulations to respond to site-specific conditions by integrating cultural, biological, and mechanical practices that foster cycling of resources, promote ecological balance, and conserve biodiversity.

For the vast majority of human history, agriculture can be described as organic; only during the 20th century was a large supply of new synthetic chemicals introduced to the food supply. This more recent style of production is referred to as "conventional." Under organic production, the use of conventional non-organic pesticide (including insecticides, fungicides, and herbicides) is precluded.

However, contrary to popular belief, certain sprays and other materials that meet organic standards are allowed in the production of organic food. If livestock are involved, the livestock must be reared with regular access to pasture and without the routine use of antibiotics or growth hormones. In most countries, organic produce may not be genetically modified. It has been suggested that the application of nanotechnology to food and agriculture is a further technology that needs to be excluded from certified organic food. The Soil Association (UK) has been the first organic certifier to implement a nano-exclusion.

Organic food production is a heavily regulated industry, distinct from private gardening. Currently, the European Union, the United

States, Canada, Japan and many other countries require producers to obtain special certification in order to market food as "organic" within their borders. Most certifications allow some chemicals and pesticides to be used, so consumers should be aware of the standards for qualifying as "organic" in their respective locales.

Historically, organic farms have been relatively small family-run operations, which is why organic food was once only available in small stores or farmers' markets. However, since the early 1990s organic food production has had growth rates of around 20% a year, far ahead of the rest of the food industry, in both developed and developing nations. As of April 2008, organic food accounts for 1–2% of food sales worldwide.

Meaning and Origin of the Term

In 1939, Lord Northbourne coined the term *organic farming* in his book *Look to the Land* (1940), out of his conception of "the farm as organism," to describe a holistic, ecologically-balanced approach to farming—in contrast to what he called *chemical farming*, which relied on "imported fertility" and "cannot be self-sufficient nor an organic whole." This is different from the scientific use of the term "organic," to refer to a class of molecules that contain carbon, especially those involved in the chemistry of life.

Environmental Impact

Several surveys and studies have attempted to examine and compare conventional and organic systems of farming. The general consensus across these surveys is that organic farming is less damaging for the following reasons:

- Organic farms do not consume or release synthetic pesticides into the environment—some of which have the potential to harm soil, water and local terrestrial and aquatic wildlife.
- Organic farms are better than conventional farms at sustaining diverse ecosystems, *i.e.*, populations of plants and insects, as well as animals.
- When calculated either per unit area or per unit of yield, organic farms use less energy and produce less waste, *e.g.*, waste such as packaging materials for chemicals.

However, some critics of organic farming methods believe that organic farms require more land to produce the same amount of food as conventional farms. They argue that if this is true, organic farms could potentially destroy the rain forests and wipe out many ecosystems.

A 2003 investigation by the Department for Environment Food and Rural Affairs in the UK found, similar to other reports, that organic farming "can produce positive environmental benefits", but that some of the benefits were decreased or lost when comparisons are made on "the basis of unit production rather than area".

Yield

One study found a 20% smaller yield from organic farms using 50% less fertilizer and 97% less pesticide. Studies comparing yields have had mixed results. Supporters claim that organically managed soil has a higher quality and higher water retention. This may help increase yields for organic farms in drought years.

One study from the Danish Environmental Protection Agency found that, area-for-area, organic farms of potatoes, sugar beet and seed grass produce as little as half the output of conventional farming. Findings like these, and the dependence of organic food on manure from low-yield cattle, has prompted criticism from scientists that organic farming is environmentally unsound and incapable of feeding the world population.

Among these critics are Norman Borlaug, father of the "green revolution," and winner of the Nobel Peace Prize, who asserts that organic farming practices can at most feed 4 billion people, after expanding cropland dramatically and destroying ecosystems in the process. Michael Pollan, author of *The Omnivore's Dilemma*, responds to this by pointing out that the average yield of world agriculture is substantially lower than modern sustainable farming yields. Bringing average world yields up to modern organic levels could increase the worlds food supply by 50 %.

A 2007 study compiling research from 293 different comparisons into a single study to assess the overall efficiency of the two agricultural systems has concluded that...organic methods could produce enough food on a global per capita basis to sustain the current human population, and potentially an even larger population, without increasing the agricultural land base. The researchers also found that while in developed countries, organic systems on average produce 92% of the yield produced by conventional agriculture, organic systems produce 80% more than conventional farms in developing countries, because the materials needed for organic farming are more accessible than synthetic farming materials to farmers in some poor countries. On the other hand, communities that lack sufficient manure to replenish soils would struggle with organic farming, and the soil would degrade rapidly.

Energy Efficiency

A study of the sustainability of apple production systems showed that in comparing a conventional farming system to an organic method of farming, the organic system is more energy efficient. However, this is debatable due to organic farming's large use of tillage for weed control. Also increased fuel use from incorporating less nutrient dense fertilizers results in higher fuel consumption rates. The general analysis is that organic production methods are usually more energy efficient because they do not use chemically synthesized nitrogen. But they generally consume more petroleum because of the lack of other options for weed control and more intensive soil management practices.

Energy efficiency is hard to determine; in the case listed above the author cites a book written in 1976. The true value of efficiency and energy consumption in relation to organic farms has yet to be determined.

Pesticides and Farmers

There are studies detailing the effects and side effects of pesticides upon the health of farm workers. Even when pesticides are organic, and are used correctly, they still end up in the air and bodies of farm workers. Through these studies, organophosphate pesticides have become associated with acute health problems such as abdominal pain, dizziness, headaches, nausea, vomiting, as well as skin and eye problems. In addition, there have been many other studies that have found pesticide exposure is associated with more severe health problems such as respiratory problems, memory disorders, dermatologic conditions, cancer, depression, neurologic deficits, miscarriages, and birth defects. Summaries of peer-reviewed research have examined the link between pesticide exposure and neurological outcomes and cancer in organophosphate-exposed workers. Those pesticides found to cause major health problems are banned for use in agriculture, conventional or organic, in many developed counties. Imported fruits and vegetables from Latin America are more likely to contain high level of pesticides, even pesticides banned for use in the United States. Migratory birds, such as Swainson's Hawks, have wintering grounds in Argentina where thousands of them were found dead from monocrotophos insecticide poisoning.

Safety and Pesticides

A study published by the National Research Council in 1993 determined that for infants and children, the major source of exposure to pesticides is through diet. A recent study in 2006 measured the levels of organophosphorus pesticide exposure in 23 schoolchildren before and after replacing their diet with organic food. In this study it was found that levels of organophosphorus pesticide exposure dropped dramatically and immediately when the children switched to an organic diet. Food residue limits established by law are set specifically with children in mind and consider a child's lifetime ingestion of each pesticide. There are controversial data on the health implications of certain pesticides.

For example, the herbicide Atrazine has been, causing demasculinization in male frogs exposed to small concentrations. Under the effects of Atrazine, male frogs were found to have greatly increased occurrences of either malformed gonads, or testicular gonads which contain non-degenerate eggs. Effects were however significantly reduced in high concentrations, as is consistent with other teratogens affecting the endocrine system, such as estradiol.

Organic farming standards do not allow the use of synthetic pesticides, but they do allow the use of specific pesticides derived from plants. The most common organic pesticides, accepted for *restricted* use by most organic standards, include Bt, pyrethrum and rotenone. Rotenone has high toxicity to fish and aquatic creatures, causes Parkinson's disease if injected into rats, and shows other toxicity to mammals.

The United States Environmental Protection Agency and state agencies periodically review the licensing of suspect pesticides, but the process of de-listing is slow. One example of this slow process is exemplified by the pesticide Dichlorvos, or DDVP, which as recently as the year 2006 the EPA proposed its continued sale. The EPA has almost banned this pesticide on several occasions since the 1970s, but it never did so despite considerable evidence that suggests DDVP is not only carcinogenic but dangerous to the human nervous system—especially in children. The EPA "has determined that risks do not exceed levels of concern", a study of longterm exposure to DDVP in rats showed no toxic effects.

A study published in 2002 showed that "Organically grown foods consistently had about one-third as many residues as conventionally grown foods." Monitoring of pesticide residues in the United States is carried out by the Pesticide Data Program, a branch of the USDA created in 1990. It has since tested over 60 different types of food for over 400 different types of pesticides – with samples collected close to the point of consumption. Their most recent results found in 2005 that:

> *"These data indicate that 29.5 percent of all samples tested contained no detectable pesticides [parent compound and metabolite(s) combined], 30 percent contained 1 pesticide, and slightly over 40 percent contained more than 1 pesticide."*
>
> —USDA, Pesticide Data Program

Several studies corroborate this finding by having found that 25 percent of organic food carries synthetic pesticide residues, in comparison to 77 percent of conventional food.

Nutritional Value and Taste

In April 2009, results from Quality Low Input Food (QLIF), a 5-year integrated study funded by the European Commission, confirmed that "the quality of crops and livestock products from organic and conventional farming systems differs considerably." Specifically, results from a QLIF project studying the effects of organic and low-input farming on crop and livestock nutritional quality "showed that organic food production methods resulted in some case: (a) higher levels of nutritionally desirable compounds (e.g., vitamins/antioxidants and poly-unsaturated fatty acids such as omega-3 and CLA); (b) lower levels of nutritionally undesirable compounds such as heavy metals, mycotoxins, pesticide residues and glyco-alkaloids in a range of crops and/or milk; (c) a lower risk of faecal Salmonella shedding in pigs." but also showed no significant difference between traditionally grown foods on other studies.

The QLIF study also concludes that "further and more detailed studies are required to provide proof for positive health impacts of organic diets on human and animal health." Alternatively, according to the UK's Food Standards Agency, "Consumers may choose to buy organic fruit, vegetables and meat because they believe them to be more nutritious than other food.

However, the balance of current scientific evidence does not support this view." A 12-month systematic review commissioned by the FSA in 2009 and conducted at the London School of Hygiene & Tropical Medicine based on 50 years' worth of collected evidence concluded that "there is no good evidence that consumption of organic food is beneficial to health in relation to nutrient content." Other studies have found no proof that organic food offers greater nutritional values, more consumer safety or any distinguishable difference in taste. A recent review of nutrition claims showed that organic food proponents are unreliable information sources which harm consumers, and that consumers are wasting their money if they buy organic food believing that that it contains better nutrients.

Regarding taste, a 2001 study concluded that organic apples were sweeter by blind taste test. Firmness of the apples was also rated higher than those grown conventionally. Current studies have not found differences in the amounts of natural biotoxins between organic and conventional foods.

Cost

Organic products typically cost 10 to 40% more than similar conventionally produced products. According to the USDA, Americans, on average, spent $1,347 on groceries in 2004; thus switching entirely to organics would raise their cost of groceries by $538.80 per year ($44.90/month) and switching to half organics would raise costs by $269.40 ($22.45/month). Processed organic foods vary in price when compared to their conventional counterparts. An Australian study by Choice magazine in 2004 found processed organic foods in supermarkets to be 65% more expensive, but noted this was not consistent. Prices may be higher because organic produce is produced on a smaller scale, and may need to be milled or processed separately. Furthermore, there is an increase in shipping costs from more centralized production in otherwise regional markets. In the case of dairy and eggs, the animal's requirements such as the number of animals that can be raised per acre, or the breed of animal and its feed conversion ratio affects the costs.

Related Movements

Biodynamic agriculture, a method of organic farming, is closely related to the organic food movement.

Food Co-ops and buyers groups are alternative ways of selling food which have also been closely related to the organic food movement. These co-operative structures have been aimed at lowering the retail price of organic food and some attempt to reduce the food miles associated with the produce.

Facts and Statistics

While organic food accounts for 1–2% of total food sales worldwide, the organic food market is growing rapidly, far ahead of the rest of the food industry, in both developed and developing nations.

- World organic food sales jumped from US $23 billion in 2002 to $52 billion in 2008.

- The world organic market has been growing by 20% a year since the early 1990s, with future growth estimates ranging from 10%–50% annually depending on the country.

Organic Clothing

Organic clothing is clothing made from materials raised or grown in compliance with organic agricultural standards.

Benefits

Authentic organic fabrics and clothing can help the environment in a number of ways, such as:

- Manufacture of chemicals is not required
- Chemical residues are not entered accidentally into the environment
- Humans and animals are not exposed to chemicals
- When the fabric is finished with chemicals are not returned to the earth in landfill, or enter into recycling process.

Cotton covers 2.5% of the world's cultivated land yet uses 16% of the world's insecticides, more than any other single major crop. It can take almost a 1/3 pound of synthetic fertilizers to grow one pound of raw cotton in the US, and it takes just under one pound of raw cotton to make one t-shirt.

Organic Cotton

Organic cotton is generally understood as cotton, from non genetically modified plants, that is to be grown without the use of any synthetic agricultural chemicals such as fertilizers or pesticides. Its production also promotes and enhances biodiversity and biological cycles. In the United States cotton plantations must also meet the requirements enforced by the National Organic Program (NOP), from the USDA, in order to be considered organic. This institution determines the allowed practices for pest control, growing, fertilizing, and handling of organic crops. As of 2007, 265,517 bales of organic cotton were produced in 24 countries and worldwide production was growing at a rate of more than 50% per year.

Ecological Footprint

Cotton covers 2.5% of the world's cultivated land yet uses 16% of the world's insecticides, more than any other single major crop. Other environmental consequences of the elevated use of chemicals in the non organic cotton growing methods consist of:

- High levels of agrochemicals are used in the production of non-organic, conventional cotton. Cotton production uses more chemicals per unit area than any other crop and accounts in total for 16% of the world's pesticides.
- Chemicals used in the processing of cotton pollute the air and surface waters.
- Residual chemicals may irritate consumers' skin.
- Decreased biodiversity and shifting equilibrium of ecosystems due to the use of pesticides.

Advantages

Cotton growers who make the transition to biologically based growing practices expect not only to offer a healthier and cleaner product, but also to benefit the planet. Some of the contributions to the different ecosystems include:

- Protecting surface and groundwater quality (eliminating contaminants in surface runoff)
- Reduced risk insect and disease control by replacing insecticide with the manipulation of ecosystems
- Long-term prevention of pests through beneficial habitat planting.
- Conservation of biodiversity
- Eliminate the use of toxic chemicals used in cotton
- Organically grown crops also yield soils with higher organic matter content, thicker topsoil depth, higher polysaccharide content, and lower modulus of rupture; therefore reducing considerably soil erosion.

Organic Certification

It is required by the law that any producer wanting to label and sell a product as "organic" must meet the standards established by the Organic Food Production Act of 1990, enforced by The State organic program (SOP) This act specifies the procedures and regulations for production and handling of organic crops.

Organic System Plan

Producers must elaborate an organic production or handling system plan which must also be approved by the state certifying agency or the USDA. This plan must include careful explanation of every process held in the plantation, as well as the frequency with which they are performed. A list of substances used on the crops is also necessary, along with a description of their composition, place where they will be used, and if possible documentation of commercial availability. This inventory of substances is important for the regulation of allowed and prohibited material established by the SOP.Organic cotton growers must also provide A description of the control procedures and physical barriers established to prevent contact of organic and non organic crops on split operations and to avoid contact of organic production with prohibited substance during gestation, harvesting, and handling operations. This production plan can also be transferred to other states as long as it has already been approved by a certifying agency.

Production

Production requirements are specifically the set of changes that must be made to field and farming practices in order for a crop to be considered organic. To begin with, organic fields must go through a cleansing period of three years, without the use of any prohibited substances, before planting the first organic crop. Fields must also be equipped with physical barriers and buzzers in order to prevent contact of organic crops with any chemical substance product of surface runoff from crops near by. Producers must also strive to promote soil fertility through cultivation practices while maintaining or improving the physical, chemical, and biological condition of the soil and minimizes soil erosion. Organic growers must also implement practices to support biodiversity. Such practices include integrated pest management (IPM), which consists of the manipulation of ecosystems that benefit both the crops and the organisms that live around it. In addition to these practices, producers may only apply crop nutrients and soil amendments included on the National List of synthetic substances allowed in crop production.

Handling

Handling procedures are all the processes related to product packaging, pest control in handling processing facilities among others. The SOP allows the use of mechanical or biological methods for the purpose of retarding spoilage of products,but at the same time it prohibits the use of volatile synthetic solvents in processed products or any ingredient that is labelled as organic.

Pesticides

Since organic cotton is grown without the use of synthetic pesticides, it should contain fewer pesticides than conventional cotton. Pesticides used in the production of conventional cotton include orthophosphates such as phorate and methamidophos, endosulfan (highly toxic to farmers, but not very environmentally persistent) and aldicarb. Other pesticides persisting in cotton fields in the United States include Trifluralin, Toxaphene and DDT. Although the last two chemicals are no longer used in the United States their long breakdown period and difficulty in removal ensures their persistence. Thus even organic cotton fields may contain them since conventional cotton fields can be transitioned to organic fields in 2–3 years.

Expanding Industry

Diverse institutions and campaigns are now educating the community about organic cotton and supporting growers on the switch to organic farming. The Sustainable Cotton Project is helping farmers in the transition from chemically dependent crops to more biological sound approaches. This institution has launched the Cleaner Cotton project, which promises to produce cotton with 73% less use of chemicals. The Community Alliance with Family Farmers (CAFF)also joined this effort since 2003. Their purpose is to educate farmers and the public about biological farming techniques and the importance of reducing chemical use in fiber and food production and supporting local farmers. In 2003, SCP joined the Community Alliance with Family Farmers (CAFF) to strengthen its operations and reach other farm and consumer audiences. CAFF and SCP provide growers with information about biological farming techniques and educate the public about the importance of reducing chemical use in fiber and food production and supporting local farmers.

Regional

Organic cotton is currently being grown successfully in many countries; the largest producers (as of 2007) are Turkey, India and China. Organic cotton production in Africa takes place in at least 8 countries. The earliest producer (1990) was the SEKEM organization in Egypt; the farmers involved later convinced the Egyptian government to convert 400,000 hectares of conventional cotton production to integrated methods, achieving a 90% reduction in the use of synthetic pesticides in Egypt and a 30% increase in yields.

Organic Baby Products

Organic baby products are made without pesticides, fertilizers, antibiotics, growth hormones, genetically modified organisms, additives, irradiation or sewage sludge. The manufacturer of such a product can only make these claims in marketing and advertising materials if the product has obtained organic certification. While organic baby products are becoming a big part of the organic movement, the upward trend has been met with some skepticism. However, organic advocacy groups such as the Organic Consumers Association work to educate the marketplace about the labelling system behind certified organic products.

Benefits

Organic baby products are substituted for conventional baby care items in order to reduce the amount of chemical exposure for rapidly developing infants.

Some physicians recommend choosing organic foods for babies from conception until the age of three. During this period of time babies' bodies and brains grow faster and they are more vulnerable to environmental toxins. The fact that babies are smaller means they absorb more pesticides per pound of body weight than adults. The bottom line is toxins stay in their systems longer and can have more impact than fully developed adults.

In addition to being free of pesticides and other additives, organic foods are believed to be more nutritional than their conventional counterparts. A 2007 European Union study confirmed that organic fruit and vegetables may have up to 40 percent more antioxidants than non-organically grown produce.

Disadvantages

Organic baby products are more expensive to produce and get certified. Therefore they are in most cases priced higher than conventional goods and are not as accessible to middle and lower income families. As a result, families that may believe there are health benefits to using organic baby products will not be able to afford them. The International Federation of Organic Agriculture Movements (IFOAM) says that while prices have been declining, there are reasons the costs of certified organic products remain higher than conventional products. They include limited supply as compared to demand, greater labour and transportation costs per unit produced. The smaller number of units involved also means marketing and distribution costs have to be spread out over a smaller number of units. As demand for organic food and products increases, technological innovations and economies of scale should reduce costs of production, processing, distribution and marketing for organic produce.

Products

Many factors are causing the accelerated growth of the organic baby product market. These include: increased frequency of baby product recalls for lead and other contaminants, more scientific studies finding toxic ingredients in baby care products, increased media exposure for the green and organic movement, the acknowledgment of the benefits of organics by some medical practitioners, and increased information sharing of the benefits in online mediums.

Baby Formula and Food

Mothers who are not breastfeeding can choose to use organic infant formula to meet their baby's nutritional requirements for the first four to six months of life. Certified organic baby formulas can be made from soy and dairy milk. Organic infant formulas can be obtained at the hospital, grocery store and through various channels on the internet.

Parents can find organic versions of baby cereals in whole grain rice and whole grain oatmeal and multi-grain varieties. Some manufacturers fortify them with iron. Most brands make organic baby food in jars. The organic foods are blended to accommodate the various textures and blends of food desired by growing babies and toddlers.

Personal Care Products

A study reported in the February 2008 issue of *Pediatrics* found babies can be exposed to toxic chemicals in ordinary personal care products such as lotions, shampoos diaper creams. While the list of chemical ingredients involved in skin care manufacturing is extensive, this particular study focused in on phthalates, a reproductive toxin that can cause damage to sexual organs and impact future fertility. Another study released in March 2008 by the Organic Consumers Association found a cancer-causing chemical 1,4 Dioxane to be present in many personal care products.

Clothing

Baby clothing companies now offer items made from organically grown fibers, such as wool, cotton, hemp, soy or bamboo, that have been grown without the use of pesticides or produced without toxic fabric finishes and flame retardants. Herbal dyed organic clothing is the perfect clothing solution for infants, juveniles,babies and kids. These herbal dyed organic cotton clothes have medicinal value attached to it which nourishes kids' skin which is very soft and prone to allergies. These herbal clothes keep your baby's skin very soft and natural. These garments are dyed using medicinal rich herbs, plants, fruits, flowers and other natural minerals which are known to have medicinal/ therapeutic properties like anti-allergic, anti-inflammatory.

Nursery Items

Mattresses and mattress pad covers should be made from organic cotton, organic wool and rubber latex from rubber-tree farms. The mattresses should be free of pesticides and other processing chemicals.

Cloth diapers made from organic cotton, hemp or wool should be free of the chemical-based super absorbent gels, pesticides and chlorine found in conventional diapers.

Sleepwear/Pajamas made from organic cotton should be free of pesticides, flame retardants, and synthetics, such as polyester and acrylic. They must be snug-fitting if they have not been treated with flame retardant chemicals.

Certified organic toys have no pesticides and contain no toxic dyes and paints. Toys should be made from organic fabrics and non-toxic stuffings; wooden toys should use water- or milk-based paints and stained with natural oils or beeswax. Upholstered nursery furniture should be made from untreated, sustainable wood and covered in organic fabrics to reduce the amount of toxins in the nursery.

Organic Fertilizer

Organic fertilizers are naturally-occurring fertilizers (e.g. peat moss or green manure), or naturally occurring mineral deposits (e.g. saltpeter).

Naturally occurring organic fertilizers include manure, slurry, worm castings, peat, seaweed, humic acid, brassin and guano. Sewage sludge use in organic agricultural operations in the U.S. has been extremely limited and rare due to USDA prohibition of the practice (due to toxic metal accumulation, among other factors).

Processed organic fertilizers include compost, bloodmeal, bone meal, humic acid, amino acids, brassin and seaweed extracts. Other examples are natural enzyme digested proteins, fish meal, and feather meal. Decomposing crop residue from prior years is another source of fertility.

Discussion of the Term 'Organic'

The notion of organic growing started out as a scientific process.

Organic systems are based on the element, Carbon. Carbon is essential to any compost process.

In later years, "organic" became a legal term.

For example, there used to be a distinction between the term "organic" and the term "pesticide free". Organic simply dealt with the use of fertilizer types. Once the term "organic" became regulated, many other factors were added. "Pesticide free" is not at all related to fertilization (plant nutrition), but has become a legal inclusion.

Likewise, in scientific terms, a fish emulsion can be a good organic fertilizer, but in some jurisdictions fish emulsion must be certified "dolphin safe" to be considered "organic".

Natural Sourcing

Animal-sourced Urea and Urea-Formaldehyde (from urine), are suitable for application organic agriculture, while pure synthetic forms are not deemed, however, pure (synthetically-produced) urea is not. The common thread that can be seen through these examples is that *organic* agriculture attempts to define itself through minimal processing (e.g. via chemical energy such as petroleum), as well as being naturally-occurring or via natural biological processes such as composting. Cover crops are also grown to enrich soil as a green manure through nitrogen fixation from the atmosphere; as well as phosphorus (through nutrient mobilization) content of soils. Powdered limestone, mined rock phosphate and Chilean saltpeter, are *inorganic* chemicals in the technical (organic chemistry) sense of the word, but are considered suitable for organic agriculture..

Advantages

Although the density of nutrients in organic material is comparatively modest, they have many advantages. The majority of nitrogen supplying organic fertilizers contain insoluble nitrogen and act as a slow-release fertilizer. By their nature, organic fertilizers increase physical and biological nutrient storage mechanisms in soils, mitigating risks of over-fertilization. Organic fertilizer nutrient content, solubility, and nutrient release rates are typically much lower than mineral (inorganic) fertilizers. A University of North Carolina study found that potential mineralizable nitrogen (PMN) in the soil was 182–285% higher in organic mulched systems, than in the synthetics control. Organic fertilizers also re-emphasize the role of humus and other organic components of soil, which are believed to play several important roles:

- Mobilizing existing soil nutrients, so that good growth is achieved with lower nutrient densities while wasting less
- Releasing nutrients at a slower, more consistent rate, helping to avoid a boom-and-bust pattern
- Helping to retain soil moisture, reducing the stress due to temporary moisture stress

- Improving the soil structure
- Helping to prevent topsoil erosion (responsible for desertfication and the Dust bowl.

Organic fertilizers also have the advantage of avoiding certain problems associated with the regular heavy use of artificial fertilizers:

- The necessity of reapplying artificial fertilizers regularly (and perhaps in increasing quantities) to maintain fertility
- Extensive runoff of soluble nitrogen and phosphorus, leading to eutrophication of bodies of water (which causes fish kills)
- Costs are lower for if fertilizer is locally available.

According to the PPI institute website, it is widely thought that organic fertilizer is better than inorganic fertilizer. However, balanced responsible use of either or both can be just as good for the soil.

Disadvantages

Organic fertilizers have the following disadvantages:

- As a dilute source of nutrients when compared to inorganic fertilizers, transporting large amount of fertilizer incurs higher costs, especially with slurry and manure.
- The composition of organic fertilizers tends to be more complex and variable than a standardized inorganic product.
- *Improperly*-processed organic fertilizers may contain pathogens from plant or animal matter that are harmful to humans or plants. However, proper composting should remove them.
- More labour is needed to compost organic fertilizer, increasing labour costs. Some of this cost is offset by reduced cash purchase.

Conventional Farming Application

In non-organic farming a compromise between the use of artificial and organic fertilizers is common, often using inorganic fertilizers supplemented with the application of organics that are readily available such as the return of crop residues or the application of manure.

Defining Large Proteins and M13 Coat Proteins

M13 is a filamentous bacteriophage composed of circular single stranded DNA (ssDNA) which is 6407 nucleotides long encapsulated in approximately 2700 copies of the major coat protein P8, and capped with 5 copies of two different minor coat proteins (P9, P6, P3) on the ends. The minor coat protein P3 attaches to the receptor at the tip of the F pilus of the host *Escherichia coli*.

Infection with filamentous phages is not lethal, however the infection causes turbid plaques in *E. coli*. It is a non-lytic virus. However a decrease in the rate of cell growth is seen in the infected cells. M13 plasmids are used for many recombinant DNA processes, and the virus has also been studied for its uses in nanostructures and nanotechnology.

Phage Particles

The phage coat is primarily assembled from a 50 amino acid protein called pVIII (or p8), which is encoded by gene VIII (or g8) in the phage genome. For a wild type M13 particle, it takes about approximately 2700 copies of p8 to make the coat about 900 nm long. The coat's dimensions are flexible though and the number of p8 copies adjusts to accommodate the size of the single stranded genome it packages. For example, when the phage genome was mutated to reduce its number of DNA bases (from 6.4 kb to 221 bp), then the number of p8 copies was decreased to fewer than 100, causing the p8 coat to shrink in order to fit the reduced genome. The phage appear to be limited at approximately twice the natural DNA content. However, deletion of a phage protein (p3) prevents full escape from the host *E. coli*, and phage that are 10-20X the normal length with several copies of the phage genome can be seen shedding from the *E. coli* host.

There are four other proteins on the phage surface, two of which have been extensively studied. At one end of the filament are five copies of the surface exposed pIX (p9) and a more buried companion protein, pVII (p7). If p8 forms the shaft of the phage, p9 and p7 form the "blunt" end that is seen in the micrographs. These proteins are very small, containing only 33 and 32 amino acids respectively, though some additional residues can be added to the N-terminal portion of each which are then presented on the outside of the coat. At the other end of the phage particle are five copies of the surface exposed pIII (p3) and its less exposed accessory protein, pVI (p6). These form the rounded tip of the phage and are the first proteins to interact with the *E. coli* host during infection. p3 is also the last point of contact with the host as new phage bud from the bacterial surface.

Phage Life-cycle

The general stages to a viral life cycle are: infection, replication of the viral genome, assembly of new viral particles and then release of the progeny particles from the host. Filamentous phage use a bacterial structure known as the F pilus to infect *E. coli*, with the M13 p3 tip contacting the TolA protein on the bacterial pilus. The phage genome is then transferred to the cytoplasm of the bacterial cell where resident proteins convert the single stranded DNA genome to a double stranded replicative form ("RF"). This DNA then serves as a template for expression of the phage genes.

Two phage gene products play critical roles in the next stage of the phage life cycle, namely amplification of the genome. pII (aka p2) nicks the double stranded form of the genome to initiate replication of the + strand. Without p2, no replication of the phage genome can occur. Host enzymes copy the replicated + strand, resulting in more copies of double stranded phage DNA. pV (aka p5) competes with double stranded DNA formation by sequestering copies of the + stranded DNA into a protein/DNA complex destined for packaging into new phage particles. Interestingly there is one additional phage-encoded protein, pX (p10), that is important for regulating the number of double stranded genomes in the bacterial host.

Without p10 no + strands can accumulate. What's particularly interesting about p10 is that it's identical to the C-terminal portion of p2 since the gene for p10 is within the gene for p2 and the protein arises from transcription initiation within gene 2. This makes the manipulation of p10 inextricably linked to manipulation of p2 (an engineering headache) but it also makes for a compact and efficient phage in nature.

Phage maturation requires the phage-encoded proteins pIV (p4), pI (p1) and its translational restart product pXI (p11). Multiple copies (on the order of 12 or 14) of p4 assemble in the outer membrane into a stable, i.e. detergent resistant, barrel-shaped structure. Similarly a handful of the p1 and p11 proteins (5 or 6 copies of each) assemble in the bacterial inner membrane, and genetic evidence suggests C-terminal portions of p1 and p11 interact with the N-terminal portion of p4 in the periplasm. Together the p1, p11, p4 complex forms channels through which mature phage are secreted from the bacterial host.

To initiate phage secretion, two of the minor phage coat proteins, p9 and p7, are thought to interact with the p5-single stranded DNA complex at a region of the DNA called the packaging sequence (aka PS). The p5 proteins covering the single stranded DNA are then replaced by p8 proteins that are embedded in the bacterial membrane and the growing phage filament is threaded through the p1, p11, p4 channel. This replacement of p5 by p8 explains the microphage data presented earlier indicate how the size of the phage particle is determined by the number of bases the phage packages.

Once the phage DNA has been fully coated with p8, the secretion terminates by adding the p3/p6 cap, and the new phage detaches from the bacterial surface. How long does all this take? Amazingly, new M13 phage particles are secreted within 10 minutes from a newly infected host and can arise at a rate of 1000/cell within the first hour of infection. The bacterial host can continue to grow and divide, allowing this process to continue indefinitely.

Replication in E. Coli

Below are steps involved with replication of M13 in *E. coli.*

- Viral (+) strand DNA enters cytoplasm
- Complementary (-) strand is synthesized by bacterial enzymes
- DNA Gyrase, a type II topoisomerase, acts on double-stranded DNA and catalyzes formation of negative supercoils in double-stranded DNA
- Final product is parental replicative form (RF) DNA
- A phage protein, pII, nicks the (+) strand in the RF
- 3'-hydroxyl acts as a primer in the creation of new viral strand
- pII circulizes displaced viral (+) strand DNA
- Pool of progeny double-stranded RF molecules produced
- Negative strand of RF is template of transcription
- mRNAs are translated into the phage proteins.

Phage proteins in the cytoplasm are pII, pX, and pV, and they are part of the replication process of DNA. The other phage proteins are synthesized and inserted into the cytoplasmic or outer membranes.

- pV dimers bind newly synthesized single-stranded DNA and prevent conversion to RF DNA
- RF DNA synthesis continues and amount of pV reaches critical concentration
- DNA replication switches to synthesis of single-stranded (+) viral DNA
- pV-DNA structures from about 800 nm long and 8 nm in diamter
- pV-DNA complex is substrate in phage assembly reaction.

Helices and Metallacycles

A helix (pl: helixes or helices) is a type of space curve, i.e. a smooth curve in three-dimensional space. It is characterized by the fact that the tangent line at any point makes a constant angle with a fixed line called the *axis*. Examples of helixes are coil springs and the handrails of spiral staircases. A "filled-in" helix – for example, a spiral ramp – is called a helicoid. Helices are important in biology, as the DNA molecule is formed as two intertwined helices, and many proteins have helical substructures, known as alpha helices. The word *helix* comes from the Greek word *ëéî*.

Types

Helices can be either right-handed or left-handed. With the line of sight along the helix's axis, if a clockwise screwing motion moves the helix away from the observer, then it is called a right-handed helix; if towards the observer then it is a left-handed helix. Handedness (or chirality) is a property of the helix, not of the perspective: a right-handed helix cannot be turned or flipped to look like a left-handed one unless it is viewed in a mirror, and vice versa.

Most hardware screws are right-handed helices. The alpha helix in biology as well as the A and B forms of DNA are also right-handed helices. The Z form of DNA is left-handed.

The pitch of a helix is the width of one complete helix turn, measured parallel to the axis of the helix.

A double helix consists of two (typically congruent) helices with the same axis, differing by a translation along the axis, which may or may not measure half the pitch.

A conic helix may be defined as a spiral on a conic surface, with the distance to the apex an exponential function of the angle indicating direction from the axis. An example is the Corkscrew roller coaster at Cedar Point amusement park.

A circular helix has constant band curvature and constant torsion.

A curve is called a general helix or cylindrical helix if its tangent makes a constant angle with a fixed line in space. A curve is a general helix if and only if the ratio of curvature to torsion is constant.

Mathematical Description

In mathematics, a helix is a curve in 3-dimensional space. The following parametrisation in Cartesian coordinates defines a helix:

$$x(t) = \cos(t),$$

$$y(t) = \sin(t).$$

As the parameter t increases, the point $(x(t),y(t),z(t))$ traces a right-handed helix of pitch 2ð and radius 1 about the z-axis, in a right-handed coordinate system.

In cylindrical coordinates (r, $è$, h), the same helix is parametrised by:

$$r(t) = 1,$$

$$\theta(t) = t,$$

$$h(t) = t.$$

A circular helix of radius a and pitch 2ðb is described by the following parametrisation:

$$x(t) = a \cos(t),$$

$$y(t) = a \sin(t),$$

$$z(t) = bt.$$

Another way of mathematically constructing a helix is to plot a complex valued exponential function taking imaginary arguments.

Except for rotations, translations, and changes of scale, all right-handed helices are equivalent to the helix defined above. The equivalent left-handed helix can be constructed in a number of ways, the simplest being to negate any one of the x, y or z components.

Arc Length, Curvature and Torsion

Examples

In music, pitch space is often modeled with helices or double helices, most often extending out of a circle such as the circle of fifths, so as to represent octave equivalency.

In organometallic chemistry, a metallacycle is a derivative of a carbocyclic compound wherein a metal has replaced at least one carbon centre. Metallacycles appear frequently as reactive intermediates in catalysis, e.g. olefin metathesis and alkyne trimerization. In organic synthesis, directed ortho metalation is widely used for the functionalization of arene rings via C-H activation. One main effect that metallic atom substitution on a cyclic carbon compound is distorting the geometry due to the large size of typical metals.

Nomenclature

A wide variety of cyclic compounds containing metals in rings are known, e.g. all chelate rings. Usually, such compounds are not classified as metallacycles, but the naming conventions are not rigidly followed. Within the area of coordination chemistry and supramolecular chemistry, examples include metallacrowns, metallacryptands, metallahelices, and molecular wheels.

Classes of Metallacycles

Metal-alkene complexes can be viewed as the smallest metallacycles, but they usually are not classified as metallacycles. In the Dewar-Chatt-Duncanson model, one resonance structure for the M(η^2-alkene) centre is the metallacyclopropane.

Metallacyclobutanes

The parent metallacyclobutane has the formula $L_nM(CH_2)_3$. A stable example is ([[triphenylphosphine | $PPh_3)_2Pt(CH_2)_3$.

Metallacyclobutane intermediates are involved in the alkene metathesis and in the oligomerization and dimerization of ethylene. Relevant to alkene metathesis, the Chauvin mechanism invokes the attack of an alkene at an electrophilic metal carbene catalyst. This work helped to validate the Chauvin mechanism for olefin metathesis.

Metallacyclopentadienes

The parent metallacyclopentadiene has the formula $L_nM(CH)_4$. Most arise from the coupling of two alkynes at a low valent metal centres such as derivatives of Co(I) and Zr(II). Late metal derivatives (Co, Ni) are intermediates in the metal-catalysed trimerization of alkynes to arenes. Early metal derivatives, i.e. derivatives of Ti and Zr, are used stoichiometrically. For example, the zirconacyclopentadiene $Cp_2ZrC_4Me_4$ is a useful carrier for $C_4Me_4^{2-}$.

Metallacyclopentanes

The parent metallacyclopentane has the formula $L_nM(CH_2)_4$. Such compounds are intermediates in the metal catalysed dimerization, trimerization, and tetramerization of ethylene to give 1-butene, 1-hexene, and 1-octene, respectively.

Metallabenzenes

The parent metallacyclobutane has the formula $L_nM(CH)_5$. Since the discovery of the structure of benzene, many analogues of this iconic structure have been described. Examples include pyridine, phosphabenzene, arsabenzene, and pyrylium. Included in this group are the metallabenzenes.

Metallabenzene complexes have been classified into three varieties; in such compounds the parent acyclic hydrocarbon ligand is viewed as the anion $C_5H_5^-$. The 6 ∂ electrons in the metallacycle conform to the Hückel (4n+2) theory.

Isolated and characterized metallabenzenes have been also been reported with metals ruthenium, iridium, platinum and rhenium.

Ortho-metallation

Metallacycles often arise by cyclization of arene-containing donor ligands, e.g. aryl phosphines and amines. An early example is the cyclization of $IrCl(PPh_3)_3$ to give the corresponding Ir(III) hydride containing a four-membered IrPCC ring. Palladium(II) and platinum(II) have long been known to ortho-metallate aromatic ligands such as azobenzene, benzylamines, and 2-phenylpyridines. These reactions are strongly influenced by substituent effects, including the Thorpe-Ingold Effect. Ligands that lack aryl substituents will sometimes cyclometalate via activation of methyl groups, an early example being the internal oxidative addition of methylphosphine ligands. Metallacycle formation interferes with intermolecular C-H activation processes. For this reason, specialized "pincer ligands" ligands have been developed that resist ortho-metallation.

Copper is a very important trace element in the human body. It takes part in a lot of biological functions such as the hemoglobin synthesis, pigmentation, bone formation, reproduction, fertility development and proper functioning of the central and peripheral nervous systems. These functions of copper are carried out in association with copper-dependent enzymes. For example, dopamine-β-hydroxylase, which is necessary for the synthesis of norepinephrine, serves for the metabolic function; the cytochrome *C* oxidase helps the cellular utilization of oxygen; the superoxide dismutase is used in the disproportionation of superoxide and soluble pyridoxal dependent monoamine oxidase assists in the oxidation of catecholamine to aldehydes.

The suitable redox potential of Cu(II) in the solution makes it play an important role in the electron transition processes in the biological systems. Furthermore, there are a number of cases where copper is needed for, or enhances the biological activity of potential ligand systems. For example, the copper(II) complexes of the *N*methylisatin β-thiosemicarbazone are much more active than the free ligand. The biological activities of other synthesized Cu(II) complexes have also been known for a long time.

Besides their biological activities, copper complexes also exhibit a number of interesting structures such as helices and metallacycles by self-assembly and recognition process. Because of its orbital configuration, Cu(II) provides a typical example of Jahn-Teller effect. Copper centres in the complexes have a variety of distortions, which have attracted the attention of chemists for a few decades.

Geometry of Cu(II) Complexes

The Cu(II) ion is the most common oxidation state of copper exists in biological systems. It is considered hard, and easily to coordinates with N and O. The coordination number is mainly four, five or six. When Cu(II) is four coordinated, the geometry of the metal centre is square planar. When Cu(II) is five coordinated, the geometry of the metal centre is square distorted pyramidal or trigonal bipyramidal. When Cu(II) is six coordinated, the geometry of the metal centre is elongated octahedron. Cu(II) may also have intermediate structures between square pyramidal and trigonal bipyramidal geometry when it is five coordinated.

The judgment on the five coordinated geometry between square pyramid and trigonal bipyramid can be made based on the τ value, which was introduced by Reedijk and co-workers. It is defined as $\tau = (\beta-\alpha)/60$, which is the ratio differences between two in-plane angles, that are larger than 90°. Normally, the τ value is between 0 and 1. The absolute number 0 represents the perfect square pyramid (*C*4v) and 1 represents the ideal trigonal bipyramidal (*D*3h). The ideal situation is seldom seen. This definition has been widely used to compare the structure of five coordinate Cu(II) complexes.

Copper (II) Complexes with Schiff Base Analogues

Considerable attention has been given to Schiff base adducts formed between pyridoxal or analogues and amino acids. Studies in pyridoxal analogues started from the interest in their biological activities. Pyridoxal is Vitamin B6, which is commonly regarded as one of the most important growth factors for both animals and humans. It was found that pyridoxal catalyses a variety of amino acid reactions. Pyridoxal 5'-phosphate, one of the most important pyridoxal analogues, acts as a principal coenzyme in almost all reactions of the primary amino acids. Many reactions are catalysed by pyridoxal 5'-phosphate containing enzymes. These include transamination, decarboxylation and racemisation. The initial combination of pyridoxal phosphate with amino acids to form a Schiff base is a common feature of these two components dependent amino acid transformations. Thus a number of pyridoxal Schiff base compounds and their analogues, with a range of amino acids have previously been synthesized.37,38 The metal complexes of Schiff base derivatives have been shown to be a useful model for more complicated metal – pyridoxal-amino acid system, which are intermediates in biologically important transmission reactions.

Texaphyrin-type Expanded Porphyrins

Texaphyrin is a sub-class of heterocyclic macrocycle molecules known as porphyrins. The molecule was invented by University of Texas at Austin professor Jonathan Sessler. The name texaphryin arose because some of the molecules have a shape that can superimpose onto the points of the star featured on the state flag of Texas. Texaphyrins were nominated as the "State Molecule of Texas", but the buckyball was chosen instead.

Sessler has described possible medicinal uses of these compounds in the *Proceedings of the National Academy of Sciences* and other scientific journals. Pharmacyclics, Inc., a publicly-traded company begun by Sessler, licensed the technology behind texaphyrins from the university in order to develop commercial and medical uses for the molecules. One is being developed under the name Antrin. Two predicted applications will be in neoplastic and cardiovascular disease.

Porphyrin

Porphyrins are a group of organic compounds of which many occur in nature. One of the best-known porphyrins is heme, the pigment in red blood cells. Heme is a cofactor of the protein hemoglobin. They are heterocyclic macrocycles composed of four modified pyrrole subunits interconnected at their α carbon atoms via methine bridges (=CH-). Porphyrins are aromatic. That is, they obey Hückel's rule for aromaticity, possessing 4n+2 = ∂ electrons (n=4 for the shortest cyclic path) that are delocalized over the macrocycle. The macrocycles, therefore, are highly-conjugated systems. As a consequence, they typically have very intense absorption bands in the visible region and may be deeply coloured; the name porphyrin comes from a Greek word for *purple*. The macrocycle has 26 pi electrons in total. The parent porphyrin is porphine, and substituted porphines are called porphyrins.

Complexes of Porphyrins and Related Molecules

Porphyrins are the conjugate acids of ligands that bind metals to form complexes. The metal ion usually has a charge of 2+ or 3+. A schematic equation for these syntheses is shown:

$$H_2\text{porphyrin} + [ML_n] \rightarrow M(\text{porphyrinate})L_{n-4} + 4\,L + 2\,H^+$$

A porphyrin without metal in its cavity is a *free base*. Some iron-containing porphyrins are called hemes. Heme-containing proteins, or *hemoproteins*, are found extensively in nature. Hemoglobin and myoglobin are two O_2-binding proteins that contain iron porphyrins.

Several other heterocycles are related to porphyrins. These include corrins, chlorins, bacteriochlorophylls, and corphins. Chlorins (2,3-dihydroporphyrin) are more reduced, contain more hydrogen than porphyrins, and feature a pyrroline subunit. This structure occurs in chlorophyll. Replacement of two of the four pyrrolic subunits with pyrrolinic subunits results in either a bacteriochlorin (as found in some photosynthetic bacteria) or an isobacteriochlorin, depending on the relative positions of the reduced rings. Some porphyrin derivatives follow Hückel's rule, but most do not.

Synthesis

Biosynthesis

The "committed step" for porphyrin biosynthesis is the formation of D-aminolevulinic acid (dALA) by the reaction of the amino acid glycine with succinyl-CoA from the citric acid cycle. Two molecules of dALA combine to give porphobilinogen (PBG), which contains a pyrrole ring. Four PBGs are then combined through deamination into hydroxymethyl bilane (HMB), which is hydrolysed to form the circular tetrapyrrole uroporphyrinogen III. This molecule undergoes a number of further modifications. Intermediates are used in different species to form particular substances, but, in humans, the main end-product protoporphyrin IX is combined with iron to form heme. Bile pigments are the breakdown products of heme.

Laboratory Synthesis

One of the more common syntheses for porphyrins is based on work by Paul Rothemund. His techniques underpin more modern syntheses such as those described by Adler and Longo. The synthesis of simple porphyrins such as *meso*-tetraphenylporphyrin (H_2TPP) is also commonly done in university teaching labs. It can be done by heating an equimolar mixture of pyrrole and benzaldehyde in refluxing propionic acid for about an hour. In this method, porphyrins are assembled from pyrrole and substituted aldehydes. Acidic conditions are essential; formic acid, acetic acid, and propionic acid are typical reaction solvents, or p-toluenesulfonic acid can be used with a non-acidic solvent. Lewis acids such as boron trifluoride etherate and ytterbium triflate have also been known to catalyse porphyrin formation. A large amount of side-product is formed and is removed, usually by recrystallization or chromatography.

Applications

Although natural porphyrin complexes are essential for life, synthetic porphyrins and their complexes have limited utility. Complexes of meso-tetraphenylporphyrin, e.g., the iron(III) chloride complex (TPPFeCl), catalyse a variety of reactions in organic synthesis, but none of them are of practical value. Porphyrin-based compounds are of interest in molecular electronics and supramolecular building blocks. Phthalocyanines, which are structurally related to porphyrins, are used in commerce as dyes and catalysts. Synthetic porphyrin dyes that are incorporated in the design of solar cells are the subject of ongoing research.

In 2008, the corporation Destiny Pharma reported successful clinical trials of an intra-nasally applied porphyrin XF-73 against methicillin-resistant *Staphylococcus aureus*.

Supramolecular Chemistry

Porphyrins are often used to construct structures in supramolecular chemistry. These systems take advantage of the Lewis acidity of the metal, typically zinc. An example of a host-guest complex that was constructed from a macrocycle composed of four porphyrins. A guest-free base porphyrin is bound to the centre by coordination with its four-pyridine substituents.

Organic Geochemistry

The field of organic geochemistry, the study of the impacts and processes that organisms have had on the Earth, had its origins in the isolation of porphyrins from petroleum. This finding helped establish the biological origins of petroleum. Petroleum is sometimes "fingerprinted" by analysis of trace amounts of nickel and vanadyl porphyrins.

Chlorophyl is a magnesium porphyrin, and heme is an iron porphyrin, but neither mineral is present in petroleum. On the other hand, nickel and vanadyl porphyrins could be related to catalytic molecules from bacteria that feed primordial hydrocarbons.

A Porphyrin-related Disease: Porphyria

Porphyrias are a group of inherited or acquired disorders of certain enzymes in the heme bio-synthetic pathway (also called porphyrin pathway). They are broadly classified as acute (hepatic) porphyrias and cutaneous (erythropoietic) porphyrias, based on the site of the overproduction and accumulation of the porphyrins (or their chemical precursors). They manifest with either neurological complications or skin problems (or occasionally both). A clinically induced and histologically identical condition is called pseudoporphyria. Pseudoporphyria is characterized by normal serum and urine porphyrin levels.

The name is likely to have been a reference to the purple discolouration of feces and urine in patients during an attack. Although original descriptions are attributed to Hippocrates, the disease was first explained biochemically by Dr Felix Hoppe-Seyler in 1874, and acute porphyrias were described by the Dutch physician Prof B.J. Stokvis in 1889.

Signs and Symptoms

Acute Porphyria

The acute, or hepatic, porphyrias primarily affect the nervous system, resulting in abdominal pain, vomiting, acute neuropathy, muscle weakness, seizures, and mental disturbances, including hallucinations, depression, anxiety, and paranoia. Cardiac arrhythmias and tachycardia (fast heart rate) may develop as the autonomic nervous system is affected. Pain can be severe and can, in some cases, be both acute and chronic in nature. Constipation is frequently present, as the nervous system of the gut is affected, but diarrhea can also occur.

Given the many presentations and the relatively uncommon occurrence of porphyria, the patient may initially be suspected to have other, unrelated conditions. For instance, the polyneuropathy of acute porphyria may be mistaken for Guillain-Barré syndrome, and porphyria testing is commonly recommended in those scenarios. Systemic lupus erythematosus features photosensitivity and pain attacks and shares various other symptoms with porphyria.

Not all porphyrias are genetic, and patients with liver disease who develop porphyria as a result of liver dysfunction may exhibit other signs of their condition, such as jaundice.

Patients with acute porphyria (AIP, HCP, VP) are at increased risk over their life for hepatocellular carcinoma (primary liver cancer) and may require monitoring. Other typical risk factors for liver cancer need not be present.

Cutaneous Porphyria

The cutaneous, or erythropoietic, porphyrias primarily affect the skin, causing photosensitivity (photodermatitis), blisters, necrosis of the skin and gums, itching, and swelling, and increased hair growth on areas such as the forehead. Often there is no abdominal pain, distinguishing it from other porphyrias.

In some forms of porphyria, accumulated heme precursors excreted in the urine may cause various changes in colour, after exposure to sunlight, to a dark reddish or dark brown colour. Even a purple hue or red urine may be seen.

Diagnosis

Porphyrin Studies

Porphyria is diagnosed through spectroscopy and biochemical analysis of blood, urine, and stool. In general, urine estimation of porphobilinogen (PBG) is the first step if acute porphyria is suspected. As a result of feedback, the decreased production of heme leads to increased production of precursors, PBG being one of the first substances in the porphyrin synthesis pathway. In nearly all cases of acute porphyria syndromes, urinary PBG is markedly elevated except for the very rare ALA dehydratase deficiency or in patients with symptoms due to hereditary tyrosinemia type I. In cases of mercury- or arsenic poisoning-induced porphyria, other changes in porphyrin profiles appear, most notably elevations of uroporphyrins I & III, coproporphyrins I & III and pre-coproporphyrin.

Repeat testing during an attack and subsequent attacks may be necessary in order to detect a porphyria, as levels may be normal or near-normal between attacks. The urine screening test has been known to fail in the initial stages of a severe life threatening attack of acute intermittent porphyria.

The bulk (up to 90%) of the genetic carriers of the more common, dominantly inherited acute hepatic porphyrias (acute intermittent porphyria, hereditary coproporphyria, variegate porphyria) have been noted in DNA tests to be latent for classic symptoms and may require DNA or enzyme testing. The exception to this may be latent post-puberty genetic carriers of hereditary coproporphyria.

As most porphyrias are rare conditions, general hospital labs typically do not have the expertise, technology or staff time to perform porphyria testing. In general, testing involves sending samples of blood, stool and urine to a reference laboratory. All samples to detect porphyrins must be handled properly. Samples should be taken during an acute attack, otherwise a false negative result may occur. Samples must be protected from light and either refrigerated or preserved. If all the porphyrin studies are negative, one has to consider pseudoporphyria. A careful medication review often will find the inciting cause of pseudoporphyria.

Additional Tests

Further diagnostic tests of affected organs may be required, such as nerve conduction studies for neuropathy or an ultrasound of the liver. Basic biochemical tests may assist in identifying liver disease, hepatocellular carcinoma, and other organ problems.

Pathogenesis

In humans, porphyrins are the main precursors of heme, an essential constituent of hemoglobin, myoglobin, catalase, peroxidase, respiratory and P450 liver cytochromes.

Deficiency in the enzymes of the porphyrin pathway leads to insufficient production of heme. Heme function plays a central role in cellular metabolism. This is not the main problem in the porphyrias; most heme synthesis enzymes—even dysfunctional enzymes—have enough residual activity to assist in heme biosynthesis. The principal problem in these deficiencies is the accumulation of porphyrins, the heme precursors, which are toxic to tissue in high concentrations. The chemical properties of these intermediates determine the location of accumulation, whether they induce photosensitivity, and whether the intermediate is excreted (in the urine or feces).

There are eight enzymes in the heme biosynthetic pathway, four of which—the first one and the last three—are in the mitochondria, while the other four are in the cytosol. Defects in any of these can lead to some form of porphyria. The hepatic porphyrias are characterized by acute neurological attacks (seizures, psychosis, extreme back and abdominal pain and an acute polyneuropathy), while the erythropoietic forms present with skin problems, usually a light-sensitive blistering rash and increased hair growth.

Variegate porphyria (also *porphyria variegata* or *mixed porphyria*), which results from a partial deficiency in PROTO oxidase, manifests itself with skin lesions similar to those of porphyria cutanea tarda combined with acute neurologic attacks. All other porphyrias are either skin- or nerve-predominant.

Porphyrin Coordinated to Iron: Heme

A heme (American English) or haem (British English) is a prosthetic group that consists of an iron atom contained in the centre of a large heterocyclic organic ring called a *porphyrin*. Not all porphyrins contain iron, but a substantial fraction of porphyrin-containing metalloproteins have heme as their prosthetic group; these are known as hemoproteins. Hemes are most commonly recognized in their presence as components of hemoglobin but they are also components of a number of other hemoproteins.

Function

Hemoproteins have diverse biological functions including the transportation of diatomic gases, chemical catalysis, diatomic gas detection, and electron transfer. The heme iron serves as a source or sink of electrons during electron transfer or redox chemistry. In peroxidase reactions, the porphyrin molecule also serves as an electron source. In the transportation or detection of diatomic gases, the gas binds to the heme iron. During the detection of diatomic gases, the binding of the gas ligand to the heme iron induces conformational changes in the surrounding protein.

It has been speculated that the original evolutionary function of hemoproteins was electron transfer in primitive sulfur-based photosynthesis pathways in ancestral cyanobacteria before the appearance of molecular oxygen.

Hemoproteins achieve their remarkable functional diversity by modifying the environment of the heme macrocycle within the protein matrix. For example, the ability of hemoglobin to effectively deliver oxygen to tissues is due to specific amino acid residues located near the heme molecule. Hemoglobin binds oxygen in the pulmonary vasculature, where the pH is high and the pCO_2 is low, and releases it in the tissues, where the situations are reversed. This phenomenon is known as the Bohr effect. The molecular mechanism behind this effect is the steric organization of the globin chain; a histidine residue, located adjacent to the heme group, becomes positively charged under acid (low pH) circumstances (which are caused by dissolved CO_2 in working muscles, etc.), sterically releasing oxygen from the heme group.

Other Hemes

Heme L is the derivative of heme B which is covalently attached to the protein of lactoperoxidase, eosinophil peroxidase and thyroid peroxidase. The addition of peroxide with the glutamyl-375 and aspartyl-225 of lactoperoxidase forms ester bonds between these amino acid residues and the heme 1- and 5-methyl groups, respectively. Similar ester bonds with these two methyl groups are thought to form in eosinophil and thyroid peroxidases. Heme L is one important characteristic of animal peroxidases; plant peroxidases incorporate heme B. Lactoperoxidase and eosinophil peroxidase are protective enzymes responsible for the destruction of invading bacteria and virus. Thyroid peroxidase is the enzyme catalysing the biosynthesis of the important thyroid hormones. Because lactoperoxidase destroys invading organisms in the lungs and excrement, it is thought to be an important protective enzyme.

Heme M is the derivative of heme B covalently bound at the active site of myeloperoxidase. Heme M also contains the two ester bonds at the heme 1- and 5-methyls, much as the other mammalian peroxidases. In addition, a unique sulfonium ion linkage between the sulfur of a methionyl aminoacid residue and the heme 2-vinyl group is formed, giving this enzyme the unique capability of easily oxidizing chloride and bromide ions. Myeloperoxidase is present in mammalian neutrophils and is responsible for the destruction of invading bacteria and virus. It also synthesizes hypobromite by “mistake” which is a known mutagenic compound.

Heme D is another derivative of heme B, but in which the propionic acid side chain at the carbon of position 6, ring III is bound to this carbon both via the usual C-C bond but also by the carboxyl oxygen, giving heme D a fifth ring and a lactone. Ring III is also hydroxylated at position 5, in a conformation trans to the new lactone group. Heme D is the site for oxygen reduction to water of many types of bacteria at low oxygen tension.

Heme S is related to heme B by the having a formyl group at position 2 in place of the 2-vinyl group. Heme S is found in the hemoglobin of marine worms. The correct structures of heme B and heme S were first elucidated by German chemist Hans Fischer.

The names of cytochromes typically (but not always) reflect the kinds of hemes they contain: cytochrome a contains heme A, cytochrome c contains heme C, etc.

Synthesis

The enzymatic process that produces heme is properly called porphyrin synthesis, as all the intermediates are tetrapyrroles that are chemically classified as porphyrins. The process is highly conserved across biology. In humans, this pathway serves almost exclusively to form heme. In other species, it also produces similar substances such as cobalamin (vitamin B_{12}).

The pathway is initiated by the synthesis of D-Aminolevulinic acid (dALA or δALA) from the amino acid glycine and succinyl-CoA from the citric acid cycle (Krebs cycle). The rate-limiting enzyme responsible for this reaction, *ALA synthase*, is strictly regulated by intracellular iron levels and heme concentration. A low-iron level, e.g., in iron deficiency, leads to decreased porphyrin synthesis, which prevents accumulation of the toxic intermediates.

This mechanism is of therapeutic importance: infusion of *heme arginate* or *hematin* can abort attacks of porphyria in patients with an inborn error of metabolism of this process, by reducing transcription of ALA synthase. The organs mainly involved in heme synthesis are the liver and the bone marrow, although every cell requires heme to function properly. Heme is seen as an intermediate molecule in catabolism of haemoglobin in the process of bilirubin metabolism.

Degradation

Degradation begins inside macrophages of the spleen, which remove old and damaged erythrocytes from the circulation. In the first step, heme is converted to biliverdin by the enzyme heme oxygenase (HOXG). NADPH is used as the reducing agent, molecular oxygen enters the reaction.

In addition, heme degradation appears to be an evolutionary conserved response to oxidative stress. Briefly, when cells are exposed to free radicals, there is a rapid induction of the expression of the stress responsive heme oxygenase-1 (Hmox1) isoenzyme that catabolizes heme. The reason why cells must increase exponentially their capability to degrade heme in response to oxidative stress remains unclear but this appears to be part of a cytoprotective response that avoids the deleterious effects of free heme.

Heme in Health and Disease

Under homeostasis, the reactivity of heme is controlled by its insertion into the "heme pockets" of hemoproteins. Under oxidative stress however, some hemoproteins, e.g. hemoglobin, can release their heme prosthetic groups. The non-protein-bound (free) heme produced in this manner becomes highly cytotoxic, most probably due to the Fe atom contained within its protoporphyrin IX ring, which can undergo Fenton chemistry to catalyse in an unfettered manner the production of free radicals. This property of free heme can sensitize a variety of cell types to undergo programmed cell death in response to pro-inflammatory agonists. This deleterious effect is thought to play an important role in the pathogenesis of certain inflammatory diseases such as malaria.

Genes

The following genes are part of the chemical pathway for making heme:

- *ALAD*: aminolevulinic acid, delta-, dehydratase
- *ALAS1*: aminolevulinate, delta-, synthase 1
- *ALAS2*: aminolevulinate, delta-, synthase 2 (sideroblastic/ hypochromic anemia)
- *CPOX*: coproporphyrinogen oxidase
- *FECH*: ferrochelatase (protoporphyria)
- *HMBS*: hydroxymethylbilane synthase
- *PPOX*: protoporphyrinogen oxidase
- *UROD*: uroporphyrinogen decarboxylase
- *UROS*: uroporphyrinogen III synthase (congenital erythropoietic porphyria).
- Porphyrin coordinated to magnesium: chlorophyll
- The one-carbon-shorter analogues: corroles, including vitamin B_{12}, which is coordinated to a cobalt
- Corphins, the highly-reduced porphyrin coordinated to nickel that binds the F430 active site in methyl coenzyme M reductase (MCR).

Methyl-coenzyme M reductase (MCR) is an enzyme that occurs in archaea and catalyzes the formation of methane by combining the hydrogen donor coenzyme B and the methyl donor coenzyme M. The enzyme has two active sites, each occupied by the nickel-containing F_{430} cofactor. This conversion is symbolically written as:

$$CH_3\text{-S-CoM} + \text{HS-CoB} \rightarrow CH_4 + \text{CoB-S-S-CoM}$$

Nitrogen-substituted Porphyrins: Phthalocyanine

Phthalocyanine is an intensely coloured macrocyclic compound that is widely used in dyeing. Phthalocyanines form coordination complexes with most elements of the periodic table. These complexes are also intensely coloured and also are used as dyes.

Properties

Phthalocyanine, abbreviated H_2Pc, has several unusual properties. It and its complexes have low solubility in virtually all solvents. One litre of warm (40 °C) benzene dissolves less than a milligram of H_2Pc and CuPc. These compounds dissolve in sulfuric acid owing to the protonation of the nitrogen centres that link the pyrrole rings.

They are also highly stable thermally and typically resist melting. CuPc sublimes at >500 °C under one atmosphere of nitrogen. Relevant to their main application, phthalocyanines strongly absorb light in the red portion of the optical spectrum (about 600 to 700 nm), thus these dyes are characteristically blue or greenish.

Phthalocyanines are structurally related to other macrocyclic pigments, especially the porphyrins. Both feature four pyrrole-like subunits linked to form a 16-membered ring. The pyrrole-like rings within H_2Pc are closely related to isoindole.

Both porphyrins and phthalocyanines function as planar tetradentate dianionic ligands that bind metals through four inwardly projecting nitrogen centres. Such complexes are formally derivatives of Pc^{2-}, the conjugate base of H_2Pc.

Many derivatives of the parent phthalocyanine are known. In addition to the ring-substituted derivatives, there also exist subphthalocyanines, superphthalocyanine, and hemiporphyrazine.

History

An unidentified blue compound, which we now know was metal-free phthalocyanine, was described in 1907. In 1927, Swiss researchers accidentally synthesized copper phthalocyanine, copper naphthalocyanine, and copper octamethylphthalocyanine in an attempted conversion of *o*-dibromobenzene into phthalonitrile. They remarked on the enormous stability of these complexes but did not further characterize these blue complexes. The same blue product was further investigated at Scottish Dyes, Ltd., Grangemouth, Scotland (later ICI). ColorantHistory.org hosts an old documentary about the discovery of the pigment.

Synthesis

Phthalocyanine forms upon heating phthalic acid derivatives that contain nitrogen functional groups. Classical precursors are phthalonitrile and diiminoisoindole. In the presence of urea, the heating of phthalanhydride is a useful route to H_2Pc. These reactions are more efficient in the presence of metal salts. Other precursors include, o-cyanobenzamide,, and phthalimide. Using such methods, approximately 57000 tons of various phthalocyanines were produced in 1985.

Applications

Approximately 25% of all artificial organic pigments are phthalocyanine derivatives. Copper phthalocyanine (CuPc) dyes are produced by introducing solubilizing groups, such as one or more sulfonic acid functions. These dyes find extensive use in various areas of textile dyeing (Direct dyes for cotton), for spin dyeing and in the paper industry. Direct blue 86 is the sodium salt of CPC-sulfonic acid whereas direct blue 199 is the quaternary ammonium salt of the CPC-sulfonic acid. The quaternary ammonium salts of these sulfonic acids used as solvent dyes because of their solubility in organic solvents, e.g. Solvent Blue 38 and Solvent Blue 48. The dye derived from cobalt phthalocyanine and an amine is Phthalogen Dye IBN. 1,3-Diiminoisoindolene, the intermediate formed during phthalocyanine manufacture, used in combination with a copper salt affords the dye GK 161.

All major artists' pigment manufacturers produce variants of copper phthalocyanine, designated colour index PB15 (blue) and colour indexes PG7 and PG36 (green).

Phthalocyanine is also commonly used as a dye in the manufacture of high-speed CD-R media. Most brands of CD-R use this dye except Taiyo Yuden and Verbatim DataLife (which use cyanine and azo dyes, respectively).

Niche Applications

Metal phthalocyanines have long been examined as catalysts for redox reactions. Areas of interest are the oxygen reduction reaction and the sweetening of gas streams by removal of hydrogen sulfide.

Phthalocyanine compounds have been investigated as donor materials in molecular electronics, e.g. organic field-effect transistors.

Porphin

Porphin, sometimes spelled porphine, is the parent chemical compound for types of biochemically significant compounds called porphyrins. The chemical formula of porphin is $C_{20}H_{14}N_4$. Porphin is an organic compound that is aromatic and heterocyclic since its chemical structure.

Structural Characteristics of Porphin

Around the perimeter of the macrocycle ring, there is a cyclic chain of sp hybridized carbon atoms, all of which are part of a conjugated double bond system, giving the molecule its aromatic character.

The aromatic character of porphin stems both from its conjugation as well as its flat or planar geometry, meaning that all the atoms lie in a single plane. Bonded to the cyclic chain of carbons are four nitrogen atoms facing the centre of the molecule, two of which are bonded to hydrogen atoms and the other two nitrogens forming part of the conjugated double bond system.

Complex with Metal in the Centre

If the two nitrogens give up their hydrogens, the four central nitrogen atoms can act as ligands to bond to a metal ion in the centre of the molecule. This is often the case with biological porphyrin compounds, especially with an iron atom in the centre. When there is no metal ion (or atom) bound to the nitrogens in the centre, then the compounds are called *free porphin* or *free porphyrins*. If they are bonded to a metal in the centre, then they are *bound*. A porphyrin with an iron atom of the type found in myoglobin, hemoglobin, or certain cytochromes is called heme.

Compounds similar to porphin are the parent compounds for similar ring systems with other central metal atoms in biochemistry. These include chlorin, which has a magnesium atom in several types of chlorophyll; bacteriochlorin, found in bacteriochlorophyll; and corrin, which has a cobalt atom in cobalamin or vitamin B12.

Chapter 6

Biomolecular Chemistry

A biomolecule is any organic molecule that is produced by a living organism, including large polymeric molecules such as proteins, polysaccharides, and nucleic acids as well as small molecules such as primary metabolites, secondary metabolites, and natural products. A more general name for this class of molecules is a biogenic substance.

As organic molecules, biomolecules consist primarily of carbon and hydrogen, nitrogen, and oxygen, and, to a smaller extent, phosphorus and sulfur. Other elements sometimes are incorporated but are much less common.

Types of Niomolecules

A diverse range of biomolecules exist, including:

- Small molecules:
  - o Lipids, phospholipids, glycolipids, sterols, glycerolipids
  - o Vitamins
  - o Hormones, neurotransmitters
  - o Metabolites
- Monomers, oligomers and polymers:

Nucleosides and Nucleotides

Nucleosides are molecules formed by attaching a nucleobase to a ribose ring. Examples of these include cytidine, uridine, adenosine, guanosine, thymidine and inosine. Nucleosides can be phosphorylated by specific kinases in the cell, producing nucleotides. Both DNA and RNA are polymers, consisting of long, linear molecules. The repeating structural units, or monomers, of the nucleic acids are called nucleotides.

Each nucleotide is made of an acyclic nitrogenous base, a pentose and one to three phosphate groups. They contain carbon, nitrogen, oxygen, hydrogen and phosphorus. They serve as sources of chemical energy (adenosine triphosphate and guanosine triphosphate), participate in cellular signalling (cyclic guanosine monophosphate and cyclic adenosine monophosphate), and are incorporated into important cofactors of enzymatic reactions (coenzyme A, flavin adenine dinucleotide, flavin mononucleotide, and nicotinamide adenine dinucleotide phosphate).

Saccharides

Monosaccharides are the simplest form of carbohydrates with only one simple sugar. They essentially contain an aldehyde or ketone group in their structure. The presence of an aldehyde group in a monosaccharide is indicated by the prefix *aldo-*. Similarly, a ketone group is denoted by the prefix *keto-*. Examples of monosaccharides are the hexoses glucose, fructose, and galactose and pentoses, ribose, and deoxyribose Consumed fructose and glucose have different rates of gastric emptying, are differentially absorbed and have different metabolic fates, providing multiple opportunities for 2 different saccharides to differentially affect food intake.

Disaccharides are formed when two monosaccharides, or two single simple sugars, form a bond with removal of water. They can be hydrolysed to yield their saccharin building blocks by boiling with dilute acid or reacting them with appropriate enzymes. Examples of disaccharides include sucrose, maltose, and lactose.

Polysaccharides are polymerized monosaccharides, complex, carbohydrates. They have multiple simple sugars. Examples are starch, cellulose, and glycogen. They are generally large and often have a complex branched connectivity.

Because of their size, polysaccharides are not water-soluble, but their many hydroxy groups become hydrated individually when exposed to water, and some polysaccharides form thick colloidal dispersions when heated in water. Shorter polysaccharides, with 2 - 10 monomers, are called oligosaccharides.

A fluorescent indicator-displacement molecular imprinting sensor was developed for discriminating saccharides. It successfully discriminated three brands of orange juice beverage. The change in fluorescence intensity of the sensing films resulting is directly related to the saccharide concentration.

Lignin

Lignin is a random polymer composed mainly of aromatic rings with short (up to three) aliphatic carbons chains connecting the rings. Lignin is the second most common biopolymer (after cellulose) and is one of the primary structural components of most plants. It contains subunits derived from *p*-coumaryl alcohol, coniferyl alcohol, and sinapyl alcohol and is unusual among biomolecules in that it is racemic i.e. it is not optically active. The lack of optical activity is because the polymerization of lignin occurs via free radical coupling reactions in which there is no preference for either configuration at a chiral centre.

Lipids

Lipids are chiefly fatty acid esters, and are the basic building blocks of biological membranes. Another biological role is energy storage (e.g., triglycerides). Most lipids consist of a polar or hydrophilic head (typically glycerol) and one to three nonpolar or hydrophobic fatty acid tails, and therefore they are amphiphilic. Fatty acids consist of unbranched chains of carbon atoms that are connected by single bonds alone (saturated fatty acids) or by both single and double bonds (unsaturated fatty acids). The chains are usually 14-24 carbon groups long, but it is always an even number.

For lipids present in biological membranes, the hydrophilic head is from one of three classes:

- Glycolipids, whose heads contain an oligosaccharide with 1-15 saccharide residues.
- Phospholipids, whose heads contain a positively charged group that is linked to the tail by a negatively charged phosphate group.
- Sterols, whose heads contain a planar steroid ring, for example, cholesterol.

Other lipids include prostaglandins and leukotrienes which are both 20-carbon fatty acyl units synthesized from arachidonic acid. They are also known as fatty acids

Amino Acids

Amino acids contain both amino and carboxylic acid functional groups. (In biochemistry, the term amino acid is used when referring to those amino acids in which the amino and carboxylate functionalities are attached to the same carbon, plus proline which is not actually an amino acid). Amino acids are the building blocks of long polymer chains.

With 2-10 amino acids such chains are called peptides, with 10-100 they are often called polypeptides, and longer chains are known as proteins. These protein structures have many structural and enzymatic roles in organisms.

There are twenty amino acids that are encoded by the standard genetic code, but there are more than 500 natural amino acids.

When amino acids other than the set of twenty are observed in proteins, this is usually the result of modification after translation (protein synthesis). Only two amino acids other than the standard twenty are known to be incorporated into proteins during translation, in certain organisms:

- Selenocysteine is incorporated into some proteins at a UGA codon, which is normally a stop codon.
- Pyrrolysine is incorporated into some proteins at a UAG codon. For instance, in some methanogens in enzymes that are used to produce methane.

Besides those used in protein synthesis, other biologically important amino acids include carnitine (used in lipid transport within a cell), ornithine, GABA and taurine.

Protein Structure

The particular series of amino acids that form a protein is known as that protein's primary structure. This sequence is determined by the genetic makeup of the individual. Proteins have several, well-classified, elements of local structure formed by intermolecular attraction, this forms the secondary structure of protein. They are broadly divided in two, alpha helix and beta sheet, also called beta pleated sheets. Alpha helices are formed of coiling of protein due to attraction between amine group of one amino acid with carboxylic acid group of other.

The coil contains about 3.6 amino acids per turn and the alkyl group of amino acid lie outside the plane of coil. Beta pleated sheets are formed by strong continuous hydrogen bond over the length of protein chain. Bonding may be parallel or antiparallel in nature. Structurally, natural silk is formed of beta pleated sheets. Usually, a protein is formed by action of both these structures in variable ratios. Coiling may also be random. The overall 3D structure of a protein is termed its tertiary structure. It is formed as result of various forces like hydrogen bonding, disulfide bridges, hydrophobic interactions, hydrophilic interactions, van der Waals force etc. When two or more different polypeptide chains cluster to form a protein, quaternary structure of protein is formed. Quaternary structure is a unique attribute of polymeric and heteromeric proteins like hemoglobin, which consists of two alpha and two beta peptide chains.

Apoenzymes

An apoenzyme is the inactive storage and generally secretory form of a protein. This is required to protect the secretory cell from the activity of that protein. Apoenzymes becomes active enzyme on addition of a cofactor. Cofactors can be either inorganic (e.g., metal ions and iron-sulfur clusters) or organic compounds, (e.g., flavin and heme). Organic cofactors can be either prosthetic groups, which are tightly bound to an enzyme, or coenzymes, which are released from the enzyme's active site during the reaction.

Isoenzymes

Isoenzymes are enzymes with similar function but different structure. They are products of different genes. They are produced in different organs to perform the same function. LDH are examples of such enzymes. Their varied levels in blood are used to determine any deformity in the organ of secretion.

Vitamins

A vitamin is a compound that is generally not synthesized by a given organism but is nonetheless vital to its survival or health (for example coenzymes). These compounds must be absorbed, or eaten, but typically only in trace quantities. When originally proposed by Casimir Funk, a Polish biochemist, he believed them to all be basic and therefore named them vital amines. The "l" was later dropped to form the word vitamines.

The field of biomolecular chemistry includes knowledge about the structure of biologically important molecules, and the relation between their structure and function. The field also includes methodology of carrying out and applying research on the structure and function of bio-molecules.

List of Biomolecules

A

For substances with an A- or α- prefix such as α-amylase, please see the parent page (in this case Amylase).

- A23187 (Calcimycin, Calcium Ionophore)
- Abamectin

- Abietic acid
- Acetic acid
- Acetylcholine
- Actin
- Actinomycin D
- Adenosine
- Adenosine diphosphate (ADP)
- Adenosine monophosphate (AMP)
- Adenosine triphosphate (ATP)
- Adenylate cyclase
- Adonitol
- Adrenaline, epinephrine
- Adrenocorticotropic hormone (ACTH)
- Aequorin
- Aflatoxin
- Agar
- Alamethicin
- Alanine
- Albumins
- Aldosterone
- Aleurone
- Alpha-amanitin
- Allantoin
- Allethrin
- α-Amanatin
- Amino acid
- Amylase
- Anabolic steroid
- Anethole
- Angiotensinogen
- Anisomycin
- Antidiuretic hormone (ADH)
- Arabinose
- Arginine

- Ascomycin
- Ascorbic acid (vitamin C)
- Asparagine
- Aspartic acid
- Asymmetric dimethylarginine
- Atrial-natriuretic peptide (ANP)
- Auxin
- Avidin
- Azadirachtin A – $C_{35}H_{44}O_{16}$

B

- Bacteriocin
- Beauvericin
- Bicuculline
- Bilirubin
- Biopolymer
- Biotin (Vitamin H)
- Brefeldin A
- Brassinolide
- Brucine

C

- Cadaverine
- Caffeine
- Calciferol (Vitamin D)
- calcitonin
- Calmodulin
- Calmodulin
- Calreticulin
- Camphor
- Cannabinol
- Capsaicin
- Carbohydrase
- Carbohydrate
- Carnitine
- Carrageenan

- Casein
- Caspase
- Cellulase
- Cellulose- $(C_6H_{10}O_5)_x$
- Cerulenin
- Cetrimonium bromide (Cetrimide)- $C_{19}H_{42}BrN$
- Chelerythrine
- Chromomycin A3
- Chaparonin
- Chitin
- α-Chloralose
- Chlorophyll
- Cholecystokinin (CCK)
- Cholesterol
- Choline
- Chondroitin sulfate
- Cinnamaldehyde
- Citral
- Citric acid
- Citrinin
- Citronellal
- Citronellol
- Citrulline
- Cobalamin (vitamin B_{12})
- Coenzyme
- Coenzyme Q
- Colchicine
- Collagen
- Coniine
- Corticosteroid
- Corticosterone
- Corticotropin-releasing hormone (CRH)
- Cortisol
- Creatine

- Creatine kinase
- Crystallin
- α-Cyclodextrin
- Cyclodextrin glycosyltransferase
- Cyclopamine
- Cyclopiazonic acid
- Cysteine
- Cystine
- Cytidine
- Cytochalasin
- Cytochalasin E
- Cytochrome
- Cytochrome C
- Cytochrome c oxidase
- Cytochrome c peroxidase
- Cytokine
- Cytosine – $C_4H_5N_3O$

D

- Deoxycholic acid
- DON (DeoxyNivalenol)
- Deoxyribofuranose
- Deoxyribose
- Deoxyribose nucleic acid (DNA)
- Dextran
- Dextrin
- DNA
- Dopamine.

E

- Enzyme
- Ephedrine
- Epinephrine – $C_9H_{13}NO_3$
- Erucic acid – $CH_3(CH_2)_7CH=CH(CH_2)_{11}COOH$
- Erythritol
- Erythropoietin (EPO)

- Estradiol
- Eugenol.

F

- Fatty acid
- Fibrin
- Fibronectin
- Folic acid (Vitamin M)
- Follicle stimulating hormone (FSH)
- Formaldehyde
- Formic acid
- Forskolin
- Fructose
- Fumonisin B1.

G

- Gamma globulin
- Galactose
- Gamma globulin
- Gamma-aminobutyric acid
- Gamma-butyrolactone
- Gamma-hydroxybutyrate (GHB)
- Gastrin
- Gelatin
- Geraniol
- Globulin
- Glucagon
- Glucosamine
- Glucose – $C_6H_{12}O_6$
- Glucose oxidase
- Gluten
- Glutamic acid
- Glutamine
- Glutathione
- Gluten
- Glycerin (glycerol)

- Glycine
- Glycogen
- Glycolic acid
- Glycoprotein
- Gonadotropin-releasing hormone (GnRH)
- Granzyme
- Green fluorescent protein
- Growth hormone
- Growth hormone-releasing hormone (GHRH)
- GTPase
- Guanine
- Guanosine
- Guanosine triphosphate (+GTP).

H

- Haptoglobin
- Hematoxylin
- Heme
- Hemerythrin
- Hemocyanin
- Hemoglobin
- Hemoprotein
- Heparan sulfate
- High density lipoprotein, HDL
- Histamine
- Histidine
- Histone
- Histone methyltransferase
- HLA antigen
- Homocysteine
- Hormone
- human chorionic gonadotropin (hCG)
- Human growth hormone
- Hyaluronate
- Hyaluronidase

- Hydrogen peroxide
- 5-Hydroxymethylcytosine
- Hydroxyproline
- 5-Hydroxytryptamine.

I

- Indigo
- Indole
- Inosine
- Inositol
- Insulin
- Insulin-like growth factor
- Integral membrane protein
- Integrase
- Integrin
- Intein
- Interferon
- Inulin
- Ionomycin
- Ionone
- Isoleucine
- Iron-sulfur cluster.

K

- K252a
- K252b
- KT5720
- KT5823
- Keratin
- Kinase.

L

For substances with an l- or L- prefix such as L-alanine or DL-alanine, please see the parent page (in this case alanine).

- Lactase
- Lactic acid
- Lactose

- Lanolin
- Lauric acid
- Leptin
- Leptomycin B
- Leucine
- Lignin
- Limonene
- Linalool
- Linoleic acid
- Linolenic acid
- Lipase
- Lipid
- Lipid anchored protein
- Lipoamide
- Lipoprotein
- Low density lipoprotein, LDL
- Luteinizing hormone (LH)
- Lycopene
- Lysine
- Lysozyme.

M

- Malic acid
- Maltose
- Melatonin
- Membrane protein
- Metalloprotein
- Metallothionein
- Methionine
- Mimosine
- Mithramycin A
- Mitomycin C
- Monomer
- Mycophenolic acid

- Myoglobin
- Myosin.

N

- Nucleic Acid.

O

- Ochratoxin A
- Oestrogens
- Oligopeptide
- Oligomycin
- Orcin
- Orexin
- Ornithine
- Oxalic acid
- Oxidase
- Oxytocin.

P

- p53
- PABA
- Paclitaxel
- Palmitic acid
- Pantothenic acid (vitamin B_5)
- parathyroid hormone (PTH)
- Paraprotein
- Pardaxin
- Parthenolide
- Patulin
- Paxilline
- Penicillic acid
- Penicillin
- Penitrem A
- Peptidase
- Pepsin
- Peptide
- Perimycin

- Peripheral membrane protein
- Perosamine
- Phenethylamine
- Phenylalanine
- Phosphagen
- phosphatase
- Phospholipid
- Phenylalanine
- Phytic acid
- Plant hormones
- Polypeptide
- Polyphenol
- Polysaccharide
- Porphyrin
- Prion
- Progesterone
- Prolactin (PRL)
- Proline
- Propionic acid
- Protamine
- Protease
- Protein
- Proteinoid
- Putrescine
- Pyrethrin
- Pyridoxine or pyridoxamine (Vitamin B6)
- Pyrrolysine
- Pyruvic acid.

Q

- Quinone.

R

- Radicicol
- Raffinose
- Renin

- Retinene
- Retinol (Vitamin A)
- Rhodopsin (visual purple)
- Riboflavin (vitamin B_2)
- Ribofuranose, Ribose
- Ribozyme
- Ricin
- RNA- Ribonucleic acid
- RuBisCO.

S

- Safrole
- Salicylaldehyde
- Salicylic acid
- Salvinorin-A – $C_{23}H_{28}O_8$
- Saponin
- Secretin
- Selenocysteine
- Selenomethionine
- Selenoprotein
- Serine
- Serine kinase
- Serotonin
- Skatole
- Signal recognition particle
- Somatostatin
- Sorbic acid
- Squalene
- Staurosporin
- Stearic acid
- Sterigmatocystin
- Sterol
- Strychnine
- Sucrose (sugar)

- Sugars (in general)
- superoxide.

T

- T2 Toxin
- Tannic acid
- Tannin
- Tartaric acid
- Taurine
- Tetrodotoxin
- Thaumatin
- Topoisomerase
- Tyrosine kinase
- Taurine
- Testosterone
- Tetrodotoxin
- Thapsigargin
- Thaumatin
- Thiamine (vitamin B_1) – $C_{12}H_{17}ClN_4OS \cdot HCl$
- Threonine
- Thrombopoietin
- Thymidine
- Thymine
- Triacsin C
- Thyroid-stimulating hormone (TSH)
- Thyrotropin-releasing hormone (TRH)
- Thyroxine (T4)
- Tocopherol (Vitamin E)
- Topoisomerase
- Triiodothyronine (T3)
- Transmembrane receptor
- Trichostatin A
- Trophic hormone
- Trypsin

- Tryptophan
- Tubulin
- Tunicamycin
- Tyrosine.

U

- Ubiquitin
- Uracil
- Urea
- Urease
- Uric acid – $C_5H_4N_4O_3$
- Uridine.

V

- Valine
- Valinomycin
- Vanabins
- Vasopressin
- Verruculogen
- Vitamins (in general)
- Vitamin A (retinol)
- Vitamin B ()
  - o Vitamin B_1 (thiamine)
  - o Vitamin B_2 (riboflavin)
  - o Vitamin B_3 (niacin or nicotinic acid)
  - o Vitamin B_4 (adenine)
  - o Vitamin B_5 (pantothenic acid)
  - o Vitamin B_6 (pyridoxine or pyridoxamine)
  - o Vitamin B_{12} (cobalamin)
- Vitamin C (ascorbic acid)
- Vitamin D (calciferol)
- Vitamin E (tocopherol)
- Vitamin F
- Vitamin H (biotin)
- Vitamin K (naphthoquinone)

- Vitamin M (folic acid)
- Vitamin P (niacin or nicotinic acid)

W

- Water
- Wortmannin

X

- Xylose

Z

- Zearalenone

Amino Acids

Amino acids are molecules containing an amine group, a carboxylic acid group and a side chain that varies between different amino acids. These molecules contain the key elements of carbon, hydrogen, oxygen, and nitrogen. These molecules are particularly important in biochemistry, where this term usually refers to alpha-amino acids with the general formula $H_2NCHRCOOH$, where R is an organic substituent.

In an alpha amino acid, the amino group is attached to the carbon atom immediately adjacent to the carboxylate group (the α–carbon), but note that other types of amino acid exist when the amino group is attached to another carbon atom (for example, in gamma-amino acids such as gamma-amino-butyric acid the carbon atom to which the amino group attaches is separated from the carboxylate group by two other carbon atoms).

The various alpha amino acids differ in which side chain (R group) is attached to their alpha carbon and these can vary in size from just one hydrogen atom in glycine or a methyl group in alanine, through to a large heterocyclic group in tryptophan.

Amino acids are critical to life, and have many functions in metabolism. One particularly important function is to serve as the building blocks of proteins, which are just linear chains of amino acids, or more precisely, amino acid residues (the remaining part of an amino acid once a water molecule has been lost due to peptide bonding). Every protein is chemically defined by the order of amino acid residues, their primary structure and this, in turn, determines their secondary structure (for example well-described features such as alpha-helices or beta-pleated sheets), tertiary structure (the shape of the individual protein, for example globular as in most enzymes or linear as in collagen) or quaternary structure (the overall shape of the protein complex if several protein monomers are grouped together, as in the hemoglobin complex, which consists of four hemoglobin monomers held together by hydrogen bonds that are capable of reacting together allosterically when one or more of them binds another molecule for example oxygen).

Just as the letters of the alphabet can be combined to form an almost endless variety of words, amino acids can be linked together in varying sequences to form a vast variety of proteins. Amino acids are also important in many other biological molecules, for example they play vital roles in coenzymes such as S-adenosylmethionine.

Due to their central role in biochemistry, amino acids are very important in nutrition and are commonly used in food technology and industry. For example, monosodium glutamate is a common flavour enhancer that gives foods the taste called *umami*. They are also used in industry where applications include the production of biodegradable plastics, drugs and chiral catalysts.

History

The first few amino acids were discovered in the early 19th century. In 1806, the French chemists Louis-Nicolas Vauquelin and Pierre Jean Robiquet isolated a compound in asparagus that proved to be asparagine, the first amino acid to be discovered.

Another amino acid that was discovered in the early 19th century was cystine, in 1810, although its monomer, cysteine, was discovered much later, in 1884. Glycine and leucine were also discovered around this time, in 1820. Usage of the term *amino acid* in the English language is from 1898.

General Structure

In the structure shown at the top of the page, R represents a side chain specific to each amino acid. The carbon atom next to the carboxyl group is called the α–carbon and amino acids with a side chain bonded to this carbon are referred to as *alpha amino acids*. These are the most common form found in nature. In the alpha amino acids, the α–carbon is a chiral carbon atom, with the exception of glycine. In amino acids that have a carbon chain attached to the α–carbon (such as lysine, shown to the right) the carbons are labelled in order as α, β, γ, δ, and so on. In some amino acids, the amine group is attached to the β or γ-carbon, and these are therefore referred to as *beta* or *gamma amino acids*.

Amino acids are usually classified by the properties of their side chain into four groups. The side chain can make an amino acid a weak acid or a weak base, and a hydrophile if the side chain is polar or a hydrophobe if it is nonpolar. The chemical structures of the twenty-two standard amino acids, along with their chemical properties, are described more fully in the article on these proteinogenic amino acids.

The phrase "branched-chain amino acids" or BCAA refers to the amino acids having aliphatic side chains that are non-linear; these are leucine, isoleucine, and valine. Proline is the only proteinogenic amino acid whose side group links to the α-amino group and, thus, is also the only proteinogenic amino acid containing a secondary amine at this position. Chemically, proline is therefore an imino acid since it lacks a primary amino group, although it is still classed as an amino acid in the current biochemical nomenclature, and may also be called an "N-alkylated alpha-amino acid".

Isomerism

Of the standard α-amino acids, all but glycine can exist in either of two optical isomers, called L or D amino acids, which are mirror images of each other. While L-amino acids represent all of the amino acids found in proteins during translation in the ribosome, D-amino acids are found in some proteins produced by enzyme posttranslational modifications after translation and translocation to the endoplasmic reticulum, as in exotic sea-dwelling organisms such as cone snails. They are also abundant components of the peptidoglycan cell walls of bacteria , and D-serine may act as a neurotransmitter in the brain. The L and D convention for amino acid configuration refers not to the optical activity of the amino acid itself, but rather to the optical activity of the isomer of glyceraldehyde from which that amino acid can theoretically be synthesized (D-glyceraldehyde is dextrorotary; L-glyceraldehyde is levorotary). Alternatively, the *(S)* and *(R)* designators are used to indicate the absolute stereochemistry. Almost all of the amino acids in proteins are *(S)* at the α carbon, with cysteine being *(R)* and glycine non-chiral. Cysteine is unusual since it has a sulfur atom at the second position in its side-chain, which has a larger atomic mass than the groups attached to the first carbon which is attached to the α-carbon in the other standard amino acids, thus the *(R)* instead of *(S)*.

Zwitterions

Amino acids have both amine and carboxylic acid functional groups and are therefore both an acid and a base at the same time. At a certain pH known as the isoelectric point an amino acid has no overall charge, since the number of protonated ammonia groups (positive charges) and deprotonated carboxylate groups (negative charges) are equal. The amino acids all have different isoelectric points. The ions produced at the isoelectric point have both positive and negative charges and are known as a *zwitterion,* which comes from the German word *Zwitter* meaning "hermaphrodite" or "hybrid". Amino acids can exist as zwitterions in solids and in polar solutions such as water, but not in the gas phase. Zwitterions have minimal solubility at their isolectric point and an amino acid can be isolated by precipitating it from water by adjusting the pH to its particular isoelectric point.

Occurrence and Functions in Biochemistry

Amino acids are the structural units that make up proteins. They join together to form short polymer chains called peptides or longer chains called either polypeptides or proteins. These polymers are linear and unbranched, with each amino acid within the chain attached to two neighbouring amino acids. The process of making proteins is called *translation* and involves the step-by-step addition of amino acids to a growing protein chain by a ribozyme that is called a ribosome. The order in which the amino acids are added is read through the genetic code from an mRNA template, which is a RNA copy of one of the organism's genes.

Twenty-two amino acids are naturally incorporated into polypeptides and are called proteinogenic or standard amino acids. Of these twenty-two, twenty are encoded by the universal genetic code. The remaining two, selenocysteine and pyrrolysine, are incorporated into proteins by unique synthetic mechanisms. Selenocysteine is incorporated when the mRNA being translated includes a SECIS element, which causes the UGA codon to encode selenocysteine instead of a stop codon. Pyrrolysine is used by some methanogenic archaea in enzymes that they use to produce methane. It is coded for with the codon UAG, which is normally a stop codon in other organisms.

Non-standard Amino Acids

Aside from the twenty-two standard amino acids, there are a vast number of "non-standard" amino acids. These non-standard amino acids found in proteins are formed by post-translational modification, which is modification after translation in protein synthesis. These modifications are often essential for the function or regulation of a protein; for example, the carboxylation of glutamate allows for better binding of calcium cations, and the hydroxylation of proline is critical for maintaining connective tissues. Another example is the formation of hypusine in the translation initiation factor EIF5A, through modification of a lysine residue. Such modifications can also determine the localization of the protein, e.g., the addition of long hydrophobic groups can cause a protein to bind to a phospholipid membrane.

Examples of nonstandard amino acids that are not found in proteins include lanthionine, 2-aminoisobutyric acid, dehydroalanine and the neurotransmitter gamma-aminobutyric acid. Nonstandard amino acids often occur as intermediates in the metabolic pathways for standard amino acids — for example ornithine and citrulline occur in the urea cycle, part of amino acid catabolism. A rare exception to the dominance of α-amino acids in biology is the β-amino acid beta alanine (3-aminopropanoic acid), which is used in plants and microorganisms in the synthesis of pantothenic acid (vitamin B_5), a component of coenzyme A.

In Human Nutrition

When taken up into the human body from the diet, the twenty-one standard amino acids are either used to synthesize proteins and other biomolecules or oxidized to urea and carbon dioxide as a source of energy. The oxidation pathway starts with the removal of the amino group by a transaminase, the amino group is then fed into the urea cycle. The other product of transamidation is a keto acid that enters the citric acid cycle. Glucogenic amino acids can also be converted into glucose, through gluconeogenesis. Pyrrolysine trait is restricted to several microbes, and only one organism has both Pyl and Sec. Of the twenty-two standard amino acids, eight are called essential amino acids because the human body cannot synthesize them from other compounds at the level needed for normal growth, so they must be obtained from food. However, the situation is quite complicated since cysteine, taurine, tyrosine, histidine and arginine are semiessential amino acids in children, because the metabolic pathways that synthesize these amino acids are not fully developed. The amounts required also depend on the age and health of the individual, so it is hard to make general statements about the dietary requirement for some amino acids.

Essential	*Nonessential*
Isoleucine	Alanine
Leucine	Asparagine
Lysine	Aspartic Acid
Methionine	Cysteine*
Phenylalanine	Glutamic Acid

Contd...

Essential	*Nonessential*
Threonine	Glutamine*
Tryptophan	Glycine*
Valine	Proline*
	Selenocysteine*
	Serine*
	Tyrosine*
	Arginine*
	Histidine*
	Ornithine*
	Taurine*

(*) Essential only in certain cases.

Non-protein Functions

In humans, non-protein amino acids also have important roles as metabolic intermediates, such as in the biosynthesis of the neurotransmitter gamma-aminobutyric acid. Many amino acids are used to synthesize other molecules, for example:

- Tryptophan is a precursor of the neurotransmitter serotonin.
- Glycine is a precursor of porphyrins such as heme.
- Arginine is a precursor of nitric oxide.
- Ornithine and S-adenosylmethionine are precursors of polyamines.
- Aspartate, glycine and glutamine are precursors of nucleotides.
- Phenylalanine is a precursor of various phenylpropanoids which are important in plant metabolism.

However, not all of the functions of other abundant non-standard amino acids are known, for example taurine is a major amino acid in muscle and brain tissues, but although many functions have been proposed, its precise role in the body has not been determined.

Some non-standard amino acids are used as defenses against herbivores in plants. For example canavanine is an analogue of arginine that is found in many legumes, and in particularly large amounts in *Canavalia gladiata* (sword bean). This amino acid protects the plants from predators such as insects and can cause illness in people if some types of legumes are eaten without processing. The non-protein amino acid mimosine is found in other species of legume, particularly *Leucaena leucocephala*. This compound is an analogue of tyrosine and can poison animals that graze on these plants.

Amino Acid Neurotransmitter

An amino acid neurotransmitter is a chemical substance which is able to transmit a nerve message across a synapse. Neurotransmitters (chemicals) are packaged into vesicles that cluster beneath the axon terminal membrane on the presynaptic side of a synapse in a process called endocytosis. Amino acid neurotransmitter release (exocytosis) is dependent upon calcium Ca and is a presynaptic response. There are inhibitory amino acids (IAA) or excitatory amino acids (EAA). Some EAA are L-Glutamate, L-Aspartate, L-Cysteine, and L-Homocysteine. These neurotransmitter systems will activate post-synaptic cells. Some IAA include GABA, Glycine, β-Alanine, and Taurine. The IAA depress the activity of post-synaptic cells.

Uses in Technology

Amino acids are used for a variety of applications in industry but their main use is as additives to animal feed. This is necessary since many of the bulk components of these feeds, such as soybeans, either have low levels or lack some of the essential amino acids: lysine, methionine, threonine, and tryptophan are most important in the production of these feeds. The food industry is also a major consumer of amino acids, particularly glutamic acid, which is used as a flavour enhancer, and Aspartame (aspartyl-phenylalanine-1-methyl ester) as a low-calorie artificial sweetener. The remaining production of amino acids is used in the synthesis of drugs and cosmetics.

Expanded Genetic Code

An expanded genetic code refers to an artificially modified genetic code in which one or more specific codons have been allocated to encode an amino acid which is not among the twenty/twentytwo found in nature. In order to understand how expansion of the genetic code is achieved, the translational mechanism catalysed by ribosomes should be considered: the central parts are the transfer RNAs (tRNA) which are used as keys to decode a three nucleotide code on the RNA being translated into its equivalent amino acid. To do this the tRNA recognises a specific codon thanks to a complementary sequence called the anticodon on one of its loops. As a result generally there are one (or more) tRNAs for each codon (or group of degenerate codons). The encoding of a codon to its amino acid is a result of the aminoacyl tRNA synthetase which adds the aminoacyl group to its allocated tRNA. However, the aminoacyl tRNA synthetase often recognises a specific motif irrespective of the anticodon, meaning that if the anticodon were to be mutated the encoding of that amino acid would change to a new codon.

For successful translation of a novel amino acid, the codon to which the amino acid is reassigned must be free or unfavoured and the novel tRNA and synthetase set (called the orthogonal set when including the codon) must not crosstalk with the endogenous tRNA and synthetase sets, while still being functionally compatible with the ribosome and other components of the translation apparatus. The tRNA synthetase pair is taken from a distant organism, generally from a different domain, and the active site of the synthetase is modified to accept the non-natural amino acid.

The possibility of reassigning codons was realized by Normanly et al. in 1990 when a viable mutant strain of E. coli read through the amber codon. As a result the amber codon became the choice codon to be assigned a novel amino acid. Later, in the Schultz lab the tRNATyr/ tyrosyl-tRNA synthetase (TyrRS) from Methanococcus jannaschii was used to introduce a tyrosine instead of STOP, the default value of the amber codon. As mentioned, this was possible because of the differences between the endogenous bacterial synthases and the orthologous archeal synthase which do not recognise each other.

Directed Evolution

This orthologous set can then be mutated and screened through directed evolution to accept a different, even novel, amino acid. Mutations to the plasmid containing the pair can be introduced by error-prone PCR or through degenerate primers for the synthetase's active site. Selection involves multiple rounds of a two-step process, where the plasmid is transferred into cells expressing chloramphenicol acetyl transferase with a premature amber codon. In the presence of toxic chloramphenicol and the non-natural amino acid, the surviving cells will have overridden the amber codon using the orthogonal tRNA aminoacylated with either the standard amino acids or the non-natural one. To remove the former, the plasmid is inserted into cells with a barnase gene (toxic) with a premature amber codon but without the non-natural amino acid, removing all the orthogonal synthases which do not specifically recognize the non-natural amino acid. In addition to the recoding of the tRNA to a different codon, they can be mutated to recognize a four base codon, allowing additional free coding options. The non natural amino acid, as a result, introduces diverse physicochemical and biological properties in order to be used as a tool to explore protein structure and function or to create novel or enhanced protein for practical purposes.

Diversity

The orthogonal pairs of synthase and tRNA which work for one organism may not work for another as the synthase may mis-aminoacylate endogenous tRNAs or the tRNA be mis-aminoacylated itself by an endogenous synthase. As a result the sets created to date differ between organisms.

Orthogonal Sets in E. coli

- $tRNA^{Tyr}$-TyrRS pair from the archaeon *Methanococcus jannaschii*
- $tRNA^{Lys}$–LysRS pair from the archaeon *Pyrococcus horikoshii*
- $tRNA^{Glu}$–GluRS pair from *Methanosarcina mazei*
- leucyl-tRNA synthetase from *Methanobacterium thermoautotrophicum* and a mutant leucyl tRNA derived from *Halobacterium* sp.

Orthogonal Sets in Yeast

- $tRNA^{Tyr}$-TyrRS pair from *Escherichia coli*
- $tRNA^{Leu}$–LeuRS pair from *Escherichia coli*
- $tRNA^{iMet}$ from human and GlnRS from *Escherichia coli.*

Orthogonal Sets in Mammalian Cells

- $tRNA^{Tyr}$-TyrRS pair from *Bacillus stearothermophilus*
- modified $tRNA^{Trp}$-TrpRS pair from *Bacillus subtilis* trp
- $tRNA^{Leu}$–LeuRS pair from *Escherichia coli.*

Protein Studies

With an expanded genetic code, the unnatural amino acid can be genetically directed to any chosen site in the protein of interest. The high efficiency and fidelity allows a better control of the placement of the modification compared to modifying the protein post-translationally, which generally will target all amino acids of the same type, such as the thiol group of cysteine and the-amino group of lysine.

Also, an expanded genetic code allows modifications to be carried out *in vivo*. The ability to site-specifically direct lab-synthesized chemical moieties into proteins allows many types of studies which would otherwise be extremely difficult.

- Probing Protein Structure and Function: by using amino acids with slightly different size such as o-Methyltyrosine or dansylalanine instead of tyrosine, and by inserting genetically-coded reporter moieties (colour-changing and/or spin-active) into selected protein sites, chemical information about the protein's structure and function can be measured.

- Identifying and Regulating Protein Activity: by using photocaged aminoacids, protein function can be "switched" on or off by illuminating the organism.
- Changing the mode of action of a protein: one can start with the gene for a protein which binds a certain sequence of DNA, and, by inserting a chemically active amino acid into the binding site, convert it to a protein which cuts the DNA, rather than binding it.
- Improving immunogenicity and overcoming self-tolerance: by replacing strategically-chosen tyrosines with *p*-nitro phenylalanine, a tolerated self-protein can be made immunogenic.

An example of the possible application for this method is the biomedical where "chemical warheads" can be added to protein which target specific cellular components.

Since 2001, 40 non-natural amino acids have been added into protein by creating a unique codon (recoding) and a corresponding transfer-RNA:aminoacyl – tRNA-synthetase pair to encode it with diverse physicochemical and biological properties in order to be used as a tool to exploring protein structure and function or to create novel or enhanced proteins.

Chemical Building Blocks

Amino acids are important as low-cost feedstocks. These compounds are used in chiral pool synthesis as enantiomerically-pure building blocks.

Amino acids have been investigated as precursors chiral catalysts, e.g. for asymmetric hydrogenation reactions, although no commercial applications exist.

Asymmetric Synthesis

Asymmetric synthesis, also called chiral synthesis, enantioselective synthesis or stereoselective synthesis, is organic synthesis that introduces one or more new and desired elements of chirality. This is important in the field of pharmaceuticals because the different enantiomers or diastereomers of a molecule often have different biological activity.

Approaches

There are three main approaches to asymmetric synthesis:

- chiral pool synthesis
- chiral auxiliaries
- asymmetric catalysis.

In practice, a mixture of all three is often used in order to maximize the advantages of each method.

Chirality must be introduced to the substance first. Then, it must be maintained. Care needs to be taken when planning the synthesis: The chirality might be removed by a chemical change that makes the substance isotropic. This process is called epimerization. For example, a S_N1 substitution reaction converts a molecule that is chiral by merit of non-planarity into a planar molecule, which has no handedness. (To visualise, draw the outlines of both of your hands on paper, and cut the images out.

You can now superimpose the images, even if the hands themselves do not superimpose.)

In a S_N2 substitution reaction, on the other hand, the chirality inverts, i.e., when you start with a right-handed mixture, you'll end up with left-handed one. (A visualization could be inverting an umbrella. The mechanism looks just the same.)

Chiral Pool Synthesis

Chiral pool synthesis is the easiest approach: A chiral starting material is manipulated through successive reactions using achiral reagents that retain its chirality to obtain the desired target molecule. This is especially attractive for target molecules having the similar chirality to a relatively inexpensive naturally occurring building-block such as a sugar or amino acid.

However, the number of possible reactions the molecule can undergo is restricted, and tortuous synthetic routes may be required. Also, this approach requires a stoichiometric amount of the enantiopure starting material, which may be rather expensive if not occurring in nature, whereas chiral catalysis requires only a catalytic amount of chiral material.

Asymmetric Induction

What many strategies in chiral synthesis have in common is asymmetric induction. The aim is to make enantiomers into diastereomers, since diastereomers have different reactivity, but enantiomers do not. To make enantiomers into diastereomers, the reagents or the catalyst need to be incorporated with an enantiopure chiral centre. The reaction will now proceed differently for different enantiomers, because the transition state of the reaction can exist in two diastereomers with respect to the enantiopure centre, and these diastereomers react differently. Asymmetric induction can also occur intramolecularly when given a chiral starting material. This chirality transfer can be exploited, especially when the goal is to make several consecutive chiral centres to give a specific enantiomer of a specific diastereomer. An aldol reaction, for example, is inherently diastereoselective; if the aldehyde is enantiopure, the resulting aldol adduct is diastereomerically and enantiomerically pure.

Chiral Auxiliary

One asymmetric induction strategy is the use of a chiral auxiliary, which forms an adduct to the starting materials and physically blocks the other trajectory for attack, leaving only the desired trajectory open. Assuming the chiral auxiliary is enantiopure, the different trajectories are not equivalent, but diastereomeric. The auxiliary shares problems similar to protecting groups; like protecting groups, auxiliaries require a reaction step to add and another to remove, increasing cost and decreasing yield.

Biocatalysis & Organocatalysis

Biocatalysis makes use of enzymes to effect chemical reagents stereoselectively. Some small organic molecules can also be used to help accelerate the desired reaction; this method is known as organocatalysis. If the organic molecule is chiral, it may react preferentially with the substrate of a certain chirality.

Asymmetric Catalysis

The oldest asymmetric synthesis is the enantioselective decarboxylation of the malonic acid *2-ethyl-2-methylmalonic acid* mediated by brucine (forming the salt) as reported by Willy Marckwald in 1904:

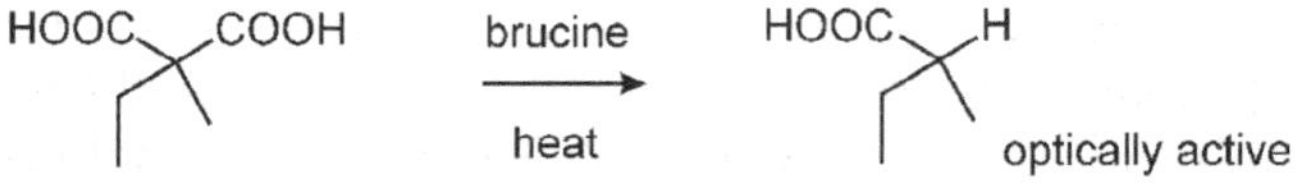

Small amounts of chiral, enantiomerically pure (or enriched) catalysts promote reactions and lead to the formation of large amounts of enantiomerically pure or enriched products. Mostly, three different kinds of chiral catalysts are employed:

1. metal ligand complexes derived from chiral ligands
2. chiral organocatalysts
3. biocatalysts.

The first methods were pioneered by William S. Knowles and Ryôji Noyori (Nobel Prize in Chemistry 2001). Knowles in 1968 replaced the achiral triphenylphosphine ligands in Wilkinson's catalyst by the chiral phosphine ligands P(Ph)(Me)(Propyl), thus creating the first asymmetric catalyst. This experimental catalyst was employed in an asymmetric hydrogenation with a modest 15% enantiomeric excess result.

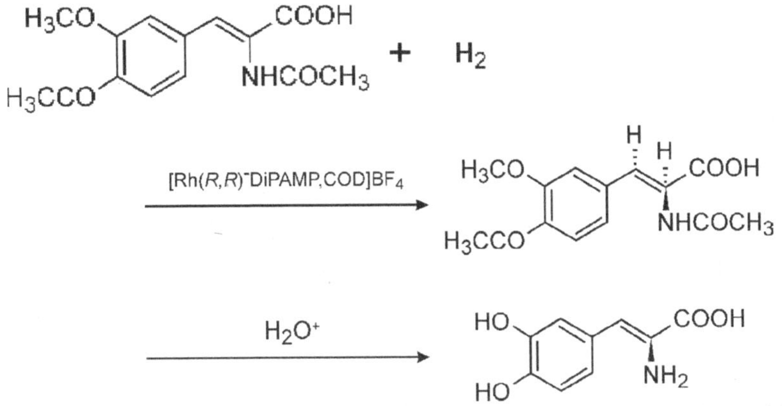

In the same year and independently, Noyori published his chiral ligand for a cyclopropanation reaction of styrene. In common with Knowles' findings, Noyori's results for the enantiomeric excess for this first-generation ligand was disappointingly low: 6%.

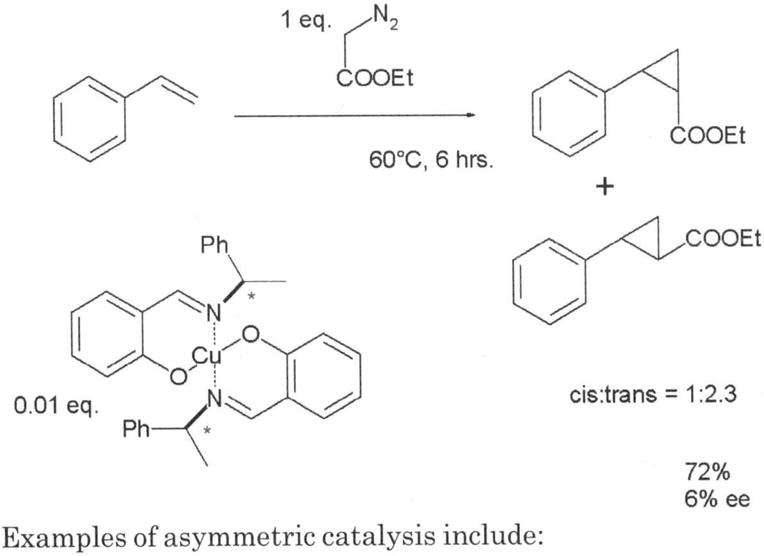

Examples of asymmetric catalysis include:

- BINAP, a chiral phosphine, used in combination with compounds of ruthenium or rhodium. These complexes catalyse the hydrogenation of functionalised alkenes well on only one face of the molecule. This process also developed by Ryôji Noyori is commercialized as the industrial synthesis of menthol using a chiral BINAP-rhodium complex.

- The other part of that Nobel prize concerned the Sharpless bishydroxylation
- Naproxen is synthesized with a chiral phosphine ligand in a hydrocyanation reaction
- asymmetric catalytic reduction and oxidation.

Alternatives

Apart from asymmetric synthesis, racemic mixtures of compounds may be separated by various techniques in chiral resolution. Where the cost in time and money of making such racemic mixtures is low, or if both enantiomers may find use, this approach may remain cost-effective.

Biodegradable Plastics

Amino acids are under development as components of a range of biodegradable polymers. These materials have applications as environmentally-friendly packaging and in medicine in drug delivery and the construction of prosthetic implants.

These polymers include polypeptides, polyamides, polyesters, polysulfides and polyurethanes with amino acids either forming part of their main chains or bonded as side chains. These modifications alter the physical properties and reactivities of the polymers. An interesting example of such materials is polyaspartate, a water-soluble biodegradable polymer that may have applications in disposable diapers and agriculture. Due to its solubility and ability to chelate metal ions, polyaspartate is also being used as a biodegradeable anti-scaling agent and a corrosion inhibitor. In addition, the aromatic amino acid tyrosine is being developed as a possible replacement for toxic phenols such as bisphenol A in the manufacture of polycarbonates.

Reactions

As amino acids have both a primary amine group and a primary carboxyl group, these chemicals can undergo most of the reactions associated with these functional groups. These include nucleophilic addition, amide bond formation and imine formation for the amine group and esterification, amide bond formation and decarboxylation for the carboxylic acid group. The multiple side chains of amino acids can also undergo chemical reactions. The types of these reactions are determined by the groups on these side chains and are therefore different between the various types of amino acid.

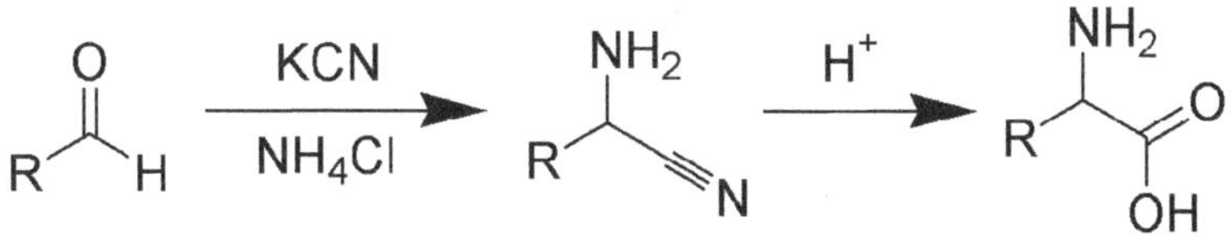

The Strecker amino acid synthesis

Chemical Synthesis

Several methods exist to synthesize amino acids. One of the oldest methods, begins with the bromination at the α-carbon of a carboxyic acid. Nucleophilic substitution with ammonia then converts the alkyl bromide to the amino acid. Alternatively, the Strecker amino acid synthesis involves the treatment of an aldehyde with potassium cyanide and ammonia, this produces an α-amino nitrile as an intermediate. Hydrolysis of the nitrile in acid then yields a α-amino acid. Using ammonia or ammonium salts in this reaction gives unsubstituted amino acids, while substituting primary and secondary amines will yield substituted amino acids. Likewise, using ketones, instead of aldehydes, gives α,α-disubstituted amino acids. The classical synthesis gives racemic mixtures of α-amino acids as products, but several alternative procedures using asymmetric auxiliaries or asymmetric catalysts have been developed.

Currently the most adopted method is an automated synthesis on a solid support (e.g. polystyrene beads), using protecting groups (e.g. Fmoc and t-Boc) and activating groups (e.g. DCC and DIC).

Peptide Bond Formation

As both the amine and carboxylic acid groups of amino acids can react to form amide bonds, one amino acid molecule can react with another and become joined through an amide linkage. This polymerization of amino acids is what creates proteins.

This condensation reaction yields the newly formed peptide bond and a molecule of water. In cells, this reaction does not occur directly; instead the amino acid is first activated by attachment to a transfer RNA molecule through an ester bond. This aminoacyl-tRNA is produced in an ATP-dependent reaction carried out by an aminoacyl tRNA synthetase. This aminoacyl-tRNA is then a substrate for the ribosome, which catalyzes the attack of the amino group of the elongating protein chain on the ester bond. As a result of this mechanism, all proteins made by ribosomes are synthesized starting at their N-terminus and moving towards their C-terminus.

However, not all peptide bonds are formed in this way. In a few cases, peptides are synthesized by specific enzymes. For example, the tripeptide glutathione is an essential part of the defenses of cells against oxidative stress. This peptide is synthesized in two steps from free amino acids. In the first step gamma-glutamylcysteine synthetase condenses cysteine and glutamic acid through a peptide bond formed between the side-chain carboxyl of the glutamate (the gamma carbon of this side chain) and the amino group of the cysteine. This dipeptide is then condensed with glycine by glutathione synthetase to form glutathione.

In chemistry, peptides are synthesized by a variety of reactions. One of the most used in solid-phase peptide synthesis, which uses the aromatic oxime derivatives of amino acids as activated units. These are added in sequence onto the growing peptide chain, which is attached to a solid resin support. The ability to easily synthesize vast numbers of different peptides by varying the types and order of amino acids (using combinatorial chemistry) has made peptide synthesis particularly important in creating libraries of peptides for use in drug discovery through high-throughput screening.

Biosynthesis and Catabolism

In plants, nitrogen is first assimilated into organic compounds in the form of glutamate, formed from alpha-ketoglutarate and ammonia in the mitochondrion. In order to form other amino acids, the plant uses transaminases to move the amino group to another alpha-keto carboxylic acid. For example, aspartate aminotransferase converts glutamate and oxaloacetate to alpha-ketoglutarate and aspartate. Other organisms use transaminases for amino acid synthesis too. Transaminases are also involved in breaking down amino acids. Degrading an amino acid often involves moving its amino group to alpha-ketoglutarate, forming glutamate. In many vertebrates, the amino group is then removed through the urea cycle and is excreted in the form of urea. However, amino acid degradation can produce uric acid or ammonia instead. For example, serine dehydratase converts serine to pyruvate and ammonia. Nonstandard amino acids are usually formed through modifications to standard amino acids. For example, homocysteine is formed through the transsulfuration pathway or by the demethylation of methionine via the intermediate metabolite S-adenosyl methionine, while hydroxyproline is made by a posttranslational modification of proline.

Microorganisms and plants can synthesize many uncommon amino acids. For example, some microbes make 2-aminoisobutyric acid and lanthionine, which is a sulfide-bridged derivative of alanine. Both of these amino acids are found in peptidic lantibiotics such as alamethicin. While in plants, 1-aminocyclopropane-1-carboxylic acid is a small disubstituted cyclic amino acid that is a key intermediate in the production of the plant hormone ethylene.

Physicochemical Properties of Amino Acids

The 20 amino acids encoded directly by the genetic code can be divided into several groups based on their properties. Important factors are charge, hydrophilicity or hydrophobicity, size and functional groups. These properties are important for protein structure and protein–protein interactions. The water-soluble proteins tend to have their hydrophobic residues (Leu, Ile, Val, Phe and Trp) buried in the middle of the protein, whereas hydrophilic side chains are exposed to the aqueous solvent. The integral membrane proteins tend to have outer rings of exposed hydrophobic amino acids that anchor them into the lipid bilayer. In the case part-way between these two extremes, some peripheral membrane proteins have a patch of hydrophobic amino acids on their surface that locks onto the membrane. Similarly, proteins that have to bind to positively-charged molecules have surfaces rich with negatively charged amino acids like glutamate and aspartate, while proteins binding to negatively-charged molecules have surfaces rich with positively charged chains like lysine and arginine. There are different hydrophobicity scales of amino acid residues.

Some amino acids have special properties such as cysteine, that can form covalent disulfide bonds to other cysteine residues, proline that forms a cycle to the polypeptide backbone, and glycine that is more flexible than other amino acids.

Many proteins undergo a range of posttranslational modifications, when additional chemical groups are attached to the amino acids in proteins. Some modifications can produce hydrophobic lipoproteins, or hydrophilic glycoproteins. These type of modification allow the reversible targeting of a protein to a membrane. For example, the addition and removal of the fatty acid palmitic acid to cysteine residues in some signalling proteins causes the proteins to attach and then detach from cell membranes.

Chemical Nature of the Amino Acids

All peptides and polypeptides are polymers of α-amino acids. There are 20 α-amino acids that are relevant to the make-up of mammalian proteins. Several other amino acids are found in the body free or in combined states (i.e. not associated with peptides or proteins). These non-protein associated amino acids perform specialized functions. Several of the amino acids found in proteins also serve functions distinct from the formation of peptides and proteins, e.g., tyrosine in the formation of thyroid hormones or glutamate acting as a neurotransmitter. The α-amino acids in peptides and proteins (excluding proline) consist of a carboxylic acid (–COOH) and an amino ($–NH_2$) functional group attached to the same tetrahedral carbon atom. This carbon is the α-carbon. Distinct R-groups, that distinguish one amino acid from another, also are attached to the alpha-carbon (except in the case of glycine where the R-group is hydrogen). The fourth substitution on the tetrahedral α-carbon of amino acids is hydrogen.

Amino Acid Classifications

Each of the 20 α-amino acids found in proteins can be distinguished by the R-group substitution on the α-carbon atom. There are two broad classes of amino acids based upon whether the R-group is hydrophobic or hydrophilic.

The hydrophobic amino acids tend to repel the aqueous environment and, therefore, reside predominantly in the interior of proteins. This class of amino acids does not ionize nor participate in the formation of H-bonds. The hydrophilic amino acids tend to interact with the aqueous environment, are often involved in the formation of H-bonds and are predominantly found on the exterior surfaces proteins or in the reactive centres of enzymes.

Acid-Base Properties of the Amino Acids

The α-COOH and $\alpha\text{-}NH_2$ groups in amino acids are capable of ionizing (as are the acidic and basic R-groups of the amino acids). As a result of their ionizability the following ionic equilibrium reactions may be written:

$$R\text{-}COOH \longleftrightarrow R\text{-}COO^- + H^+$$

$$R\text{-}NH_3^+ \longleftrightarrow R\text{-}NH_2 + H^+$$

The equilibrium reactions, as written, demonstrate that amino acids contain at least two weakly acidic groups. However, the carboxyl group is a far stronger acid than the amino group. At physiological pH (around 7.4) the carboxyl group will be unprotonated and the amino group will be protonated. An amino acid with no ionizable R-group would be electrically neutral at this pH. This species is termed a zwitterion.

Like typical organic acids, the acidic strength of the carboxyl, amino and ionizable R-groups in amino acids can be defined by the association constant, K_a or more commonly the negative logrithm of K_a, the pK_a. The net charge (the algebraic sum of all the charged groups present) of any amino acid, peptide or protein, will depend upon the pH of the surrounding aqueous environment. As the pH of a solution of an amino acid or protein changes so too does the net charge. This phenomenon can be observed during the titration of any amino acid or protein. When the net charge of an amino acid or protein is zero the pH will be equivalent to the isoelectric point: pI.

Functional Significance of Amino Acid R-Groups

In solution it is the nature of the amino acid R-groups that dictate structure-function relationships of peptides and proteins. The hydrophobic amino acids will generally be encountered in the interior of proteins shielded from direct contact with water. Conversely, the hydrophilic amino acids are generally found on the exterior of proteins as well as in the active centres of enzymatically active proteins. Indeed, it is the very nature of certain amino acid R-groups that allow enzyme reactions to occur.

The imidazole ring of histidine allows it to act as either a proton donor or acceptor at physiological pH. Hence, it is frequently found in the reactive centre of enzymes. Equally important is the ability of histidines in hemoglobin to buffer the H^+ ions from carbonic acid ionization in red blood cells. It is this property of hemoglobin that allows it to exchange O_2 and CO_2 at the tissues or lungs, respectively.

The primary alcohol of serine and threonine as well as the thiol (–SH) of cysteine allow these amino acids to act as nucleophiles during enzymatic catalysis. Additionally, the thiol of cysteine is able to form a disulfide bond with other cysteines:

Cysteine-SH + HS-Cysteine <——> Cysteine-S-S-Cysteine

This simple disulfide is identified as cystine. The formation of disulfide bonds between cysteines present within proteins is important to the formation of active structural domains in a large number of proteins. Disulfide bonding between cysteines in different polypeptide chains of oligomeric proteins plays a crucial role in ordering the structure of complex proteins, e.g. the insulin receptor.

Optical Properties of the Amino Acids

A tetrahedral carbon atom with 4 distinct constituents is said to be chiral. The one amino acid not exhibiting chirality is glycine since its "'R-group" is a hydrogen atom. Chirality describes the handedness of a molecule that is observable by the ability of a molecule to rotate the plane of polarized light either to the right (dextrorotatory) or to the left (levorotatory). All of the amino acids in proteins exhibit the same absolute steric configuration as L-glyceraldehyde. Therefore, they are all L-α-amino acids. D-amino acids are never found in proteins, although they exist in nature. D-amino acids are often found in polypetide antibiotics.

The aromatic R-groups in amino acids absorb ultraviolet light with an absorbance maximum in the range of 280nm. The ability of proteins to absorb ultraviolet light is predominantly due to the presence of the tryptophan which strongly absorbs ultraviolet light.

The Peptide Bond

Peptide bond formation is a condensation reaction leading to the polymerization of amino acids into peptides and proteins. Peptides are small consisting of few amino acids. A number of hormones and neurotransmitters are peptides. Additionally, several antibiotics and antitumor agents are peptides. Proteins are polypeptides of greatly divergent length. The simplest peptide, a dipeptide, contains a single peptide bond formed by the condensation of the carboxyl group of one amino acid with the amino group of the second with the concomitant elimination of water. The presence of the carbonyl group in a peptide bond allows electron resonance stabilization to occur such that the peptide bond exhibits rigidity not unlike the typical –C=C– double bond. The peptide bond is, therefore, said to have partial double-bond character.

Bibliography

Ausubel, F.M. : *Current Protocols in Molecular Biology,* New York: John Wiley and Sons, 1989.

Barrington, E. J. W.: *Biochemistry of Primitive Deuterostomians*, London, Academic Press, 1974.

Bauer, M.W., and G. Gaskell : *Biotechnology: The Making of a Global Controversy*, Cambridge: Cambridge University Press, 2002.

Broach, J.R. : *The Molecular Biology of the Yeast,* Cold Spring Harbor: Cold Spring Harbor Laboratory, 1981.

Bruce, Donald : *Engineering Genesis: The Ethics of Genetic Engineering in Non-Human Animals*, London: Earthscan, 1998.

Cahill, Lisa : *Genetics, Theology, and Ethics: An Interdisciplinary Conversation*, New York: Crossroad, 2005.

Carroll SB, Grenier J, Weatherbee SD: *From DNA to Diversity: Molecular Genetics and the Evolution of Animal Design*. Oxford: Blackwell Publishing, 2005.

Chambers, K. L.: *Biochemical Coevolution: Eugene,* Oregon State University Press, 1970.

Clark, J.M. and Switzer, R.L. : *Experimental Biochemistry,* New York: W.H. Freeman and Company, 1977.

David M.: *Basic Principles and Calculations in Chemical Engineering*, Prentice-Hall, 1996.

Drlica, Karl : *Understanding DNA and Gene Cloning: A Guide for the Curious*, New York: Wiley, 1997.

Fumento, Michael : *Bioevolution: How Biotechnology is Changing Our World*, San Francisco: Encounter, 2003.

Ganguly, Smriti : *Biochemistry of Biomolecules,* Pearl Books, Delhi, 2007.

Goswami, Chanderlekha: *Biochemistry and Instrumentation*, Manglam Pub, Delhi, 2011.

Haber, L. F. : *The Chemical Industry: 1900-1930, International Growth and Technological Change*, Oxford: Claredon Press, 1971.

Jack J. Pasternak : *Molecular Biotechnology: Principles and Applications of Recombinant DNA*, Washington, DC: ASM Press, 1998.

Kumar , Naren : *Fundamentals of Biochemistry : A Practical Approach,* Kanishka, Delhi, 2005.

Lodish, Harvey : *Molecular Cell Biology*, New York: W. H. Freeman, 2000.

Mahadev Sharma and Nirmal Tripathi : *Molecular Biotechnology,* Crescent Publishing Corporation, Delhi, 2008.

Morholt, E. and Brandwein, P.F. : *Sourcebook for the Biological Sciences,* San Diego: Harcourt Brace Jovanovich, 1986.

Morris, Jonathan : *Ethics and Biotechnology*, New York: Chelsea House, 2005.

Narvekar, Raghunath : *Molecular Biochemistry : Principles and Practices,* Adhyayan Pub, Delhi, 2008.

Nwabueze, Remigius N. : *Biotechnology and the Challenge of Property,* Burlington, VT: Ashgate, 2007.

Patel, S V ; U K Kandoliya: *Biochemistry at a Glance*, Agrotech Pub, Delhi, 2007.

Patrick Salomon: *A Handbook on Nanochemistry*, Dominant, Delhi, 2008.

Prasad, S.K. : *Biochemistry of Carbohydrates,* Discovery Pub, Delhi, 2010.

Reuter, Lars : *Modern Biotechnology in Postmodern Times? A Reflection on European Policies and Human Agency,* Dordrecht, Netherlands: Springer, 2003.

Richard Luxton: *Clinical Biochemistry*, Viva Books, Delhi, 2010.

Robin Whyman: *Applied Organometallic Chemistry and Catalysis*, Oxford University Press, Delhi, 2005.

Sarath Chandra Bose: *Biochemistry : A Practical Manual*, Pharma Med Press, Delhi, 2010.

Suri, Nitin: *Molecular Biology and Biochemistry,* Oxford Book Company, Delhi, 2010.

Swarnim, K. : *A Textbook of Biochemistry and Microbiology*, Surendra Pub, Delhi, 2010.

Thiele, Felix, and R.E. Ashcroft : *Bioethics in a Small World*, Berlin: Springer, 2004.

Tripathi, Nirmal : *Molecular Biotechnology,* Crescent Publishing Corporation, Delhi, 2008.

Waters, Brent & Ronald Cole-Turner : *God and the Embryo: Religious Voices on Stem Cells and Cloning*, Washington, DC: Georgetown University Press, 2003.

Yadav, M : *Nutritional Biochemistry and Metabolism*, Arise Pub, Delhi, 2008.

Index

H

I

K

L

M

N

O

P

R

S

T

V

W

Z

❑❑❑